Nadiar M. Mussin • Amin Tamadon

Mesenchymal Stem Cells and Their Extracellular Vesicles

A Translational Roadmap from Biology to Clinical Therapies

Springer

Nadiar M. Mussin
Department of Surgery
West Kazakhstan Marat Ospanov Medical University
Aktobe, Kazakhstan

Amin Tamadon
Department of Natural Sciences
West Kazakhstan Marat Ospanov Medical University
Aktobe, Kazakhstan

ISBN 978-3-032-36436-4 ISBN 978-3-032-36437-1 (eBook)
https://doi.org/10.1007/978-3-032-36437-1

This Springer imprint is published by the registered company Springer Nature Switzerland AG
The registered company address is: Gewerbestrasse 11, 6330 Cham, Switzerland

Preface

Regenerative medicine has undergone a profound conceptual transformation over the past two decades. Initially driven by the promise of direct cell replacement, the field has increasingly shifted toward understanding cells as dynamic biological factories capable of modulating tissue repair, immune responses, and microenvironmental homeostasis. Among the various cellular platforms explored, mesenchymal stromal/stem cells (MSCs) have emerged as one of the most clinically advanced and biologically versatile systems. More recently, their secreted extracellular vesicles (EVs) have gained significant attention as a cell-free therapeutic alternative, offering unique advantages in safety, scalability, and regulatory feasibility.

This book was conceived to provide a comprehensive and translationally oriented synthesis of MSC and MSC-derived EV biology, bridging fundamental mechanisms with clinical application and industrial development. While numerous reviews and articles exist in this rapidly evolving field, there remains a critical need for an integrated resource that connects molecular biology, experimental evidence, manufacturing considerations, regulatory frameworks, and future therapeutic strategies within a single coherent narrative.

The primary aim of this book is to serve as a roadmap for researchers, clinicians, and industry professionals navigating the transition from experimental discovery to clinical implementation. Each chapter is structured to progressively build understanding—from stem cell fundamentals and MSC biology to EV biogenesis, functional mechanisms, delivery strategies, preclinical evidence, and ultimately regulatory and commercialization pathways. Particular emphasis has been placed on methodological rigor, reproducibility standards, and the emerging concept of EVs as programmable biologics.

The intended audience includes graduate students, biomedical researchers, clinicians in regenerative medicine, and professionals in biotechnology and pharmaceutical development. By combining mechanistic insights with translational perspectives, this book seeks not only to summarize the current state of the field but also to highlight unresolved challenges and define priorities for future research. We hope this work contributes to advancing the scientific and clinical development of MSC- and EV-based therapies and supports the realization of precision regenerative medicine.

Aktobe, Kazakhstan

Nadiar M. Mussin
Amin Tamadon

Acknowledgments

The authors would like to express their sincere appreciation to colleagues, collaborators, and institutions who have contributed to the development of this work. Special thanks are extended to the research teams and academic partners involved in ongoing projects related to stem cell biology and extracellular vesicle therapeutics.

The authors would like to specifically acknowledge the valuable contributions and support of Kulyash Zhilisbayeva, Dr. Madina A. Kurmanalina, Dr. Akmaral Baspakova, Dr. Mahdi Mahdipour, Prof. Dr. Nader Tanideh, Dr. Afshin Zare, and Prof. Dr. Ogay V. Borisovich, whose assistance and collaboration have been instrumental in advancing related research activities.

We also acknowledge the support of West Kazakhstan Marat Ospanov Medical University for providing an academic environment conducive to advanced research and scientific writing.

The authors would also like to express their sincere appreciation to the administration of West Kazakhstan Marat Ospanov Medical University for their continuous support, leadership, and commitment to advancing academic and scientific excellence. We are particularly grateful to Talgar Satybaevich Abilov, Chairman of the Board—Rector; Nurgul Abdullaeva Abenova, Vice-Rector for Academic and Educational Work; Yerbolat Maratovich Iztleuov, Vice-Rector for Strategic Development, Project Management, International Cooperation and Digitalization; and Erlan Shuenishbaevich Zhienalin, Executive Director, for their strategic guidance and institutional support, which have played an essential role in facilitating the successful completion of this work.

Funding This book was supported by the program-targeted financing on scientific programs of the Ministry of Healthcare of the Republic of Kazakhstan, titled "Development of an exosome isolation kit from umbilical cord mesenchymal stem cell culture for therapeutic and research application" (2024–2026) (IRN BR25593457).

The authors are grateful to all researchers whose work has contributed to the evolving understanding of mesenchymal stromal cells and extracellular vesicles, forming the scientific foundation upon which this book is built.

Contents

About the Authors

Nadiar M. Mussin is an Associate Professor in Transplant Surgery at West Kazakhstan Marat Ospanov Medical University, Kazakhstan, and a researcher in regenerative medicine with a primary focus on extracellular vesicles (EVs), mesenchymal stromal/stem cells (MSCs), and cell-free therapeutic strategies. His work integrates molecular and cellular biology with translational research approaches aimed at bridging experimental discoveries and clinical applications.

He has been actively involved in studies investigating EV biogenesis, cargo profiling, and therapeutic mechanisms across multiple disease models, including inflammatory and degenerative conditions. His research interests also encompass advanced characterization techniques, bioengineering of EVs, and the optimization of delivery strategies for targeted therapies. Through his contributions to multidisciplinary research projects, he has developed strong expertise in experimental design, data analysis, and scientific writing, particularly in the fields of regenerative medicine and biomedical innovation.

Amin Tamadon is a Professor of Stem Cell Research and Principal Investigator at West Kazakhstan Marat Ospanov Medical University, Kazakhstan. He graduated from Shiraz University, Iran. He advanced his research expertise through postdoctoral fellowships at Seoul National University, South Korea, and Fudan University, China.

Prof. Dr. Tamadon's research focuses on mesenchymal stem cells, extracellular vesicles, and their applications in regenerative medicine, disease modeling, and translational therapeutics. He has extensive experience in both preclinical and clinical research, including animal models, biomaterials, and advanced cellular therapies. His work also encompasses systematic reviews, meta-analyses, and bibliometric analyses in biomedical sciences.

He is the founder of PerciaVista R&D Co., where he leads research and development initiatives in stem cell therapy and exosome-based technologies, including the development of innovative exosome isolation platforms. In addition to his research activities, he plays an active role in academic publishing and scientific communication, contributing to multiple peer-reviewed journals and international collaborations. His multidisciplinary expertise bridges basic science, clinical translation, and industrial application in regenerative medicine.

Abbreviations

ATMP	Advanced Therapy Medicinal Product
BD	Biodistribution
CM	Conditioned Medium
CQA	Critical Quality Attribute
EMA	European Medicines Agency
EV	Extracellular Vesicle
FDA	Food and Drug Administration
GMP	Good Manufacturing Practice
miRNA	MicroRNA
MoA	Mechanism of Action
MSC	Mesenchymal Stromal/Stem Cell
NTA	Nanoparticle Tracking Analysis
PBS	Phosphate-Buffered Saline
PK	Pharmacokinetics
RNA	Ribonucleic Acid
SEC	Size Exclusion Chromatography
sEV	Small Extracellular Vesicle
TFF	Tangential Flow Filtration
UC	Ultracentrifugation

Declaration of AI-Assisted Manuscript Preparation

During the preparation of this book, the authors used ChatGPT (OpenAI; Pro 5.6 Sol) to assist with the organization, drafting, and linguistic refinement of selected sections of the manuscript. All AI-assisted text was critically reviewed and revised by the authors, and the scientific statements and references were checked against the original sources. The authors exercised independent scholarly judgment in evaluating the evidence and determining the interpretations and conclusions presented in the book. The authors take full responsibility for the accuracy, originality, and integrity of the final published content.

Chapter 1
Stem Cells in Regenerative Medicine: From Cell Therapy to Cell-Free Therapy

Abstract This chapter provides a comprehensive overview of the evolution of regenerative medicine from cell-based therapies to cell-free therapeutic strategies. It introduces the fundamental classification of stem cells and explains the emergence of mesenchymal stromal/stem cells (MSCs) as a dominant platform due to their immunomodulatory and paracrine properties. The chapter further explores the role of extracellular vesicles (EVs) as key mediators of MSC therapeutic effects and discusses the transition from experimental research to clinical translation. Key challenges, regulatory developments, and future directions in MSC- and EV-based therapies are also highlighted.

Keywords Stem cells · Regenerative medicine · Mesenchymal stromal/stem cells · MSC therapy · Extracellular vesicles · Cell-free therapy · Pluripotent stem cells · Paracrine signaling · Immunomodulation · Potency assays · GMP manufacturing · Clinical translation · Biodistribution · Regulatory science · Biologic development

Regenerative medicine has moved from early expectations of cell replacement toward a more mechanistic view in which many clinically relevant stem cell interventions act as context-dependent biologic drugs, reshaping immunity and tissue repair through secreted factors rather than durable engraftment. This mechanistic pivot is central to understanding why mesenchymal stromal/stem cells (MSCs) rose to dominate adult stem cell development programs and why extracellular vesicles (EVs) are now being developed as "cell-free" therapeutic products [1].

Within the last decade, pluripotent stem cell (PSC) translation has accelerated with increasing numbers of human pluripotent stem cell (hPSC) products in interventional trials and early clinical signals across ophthalmology and neurology; a 2025 landscape review identified 83 hPSC-derived products tested in 115 regulatory authority–approved clinical trials, with more than 1200 patients dosed and no consistent class-wide safety signal reported to date, although product-specific risks—including tumorigenicity, immune responses, and ectopic tissue formation—remain important considerations requiring long-term follow-up [2].

In parallel, MSC therapy has progressed from thousands of treated patients and heterogeneous efficacy signals to a major regulatory milestone: in December 2024,

N. M. Mussin, A. Tamadon, *Mesenchymal Stem Cells and Their Extracellular Vesicles*, https://doi.org/10.1007/978-3-032-36437-1_1

the US Food and Drug Administration [3] approved remestemcel-L-rknd (Ryoncil) for pediatric steroid-refractory acute graft-versus-host disease (SR-aGVHD), marking the first MSC therapy approval in the United States [4].

The "cell-free" turn is not a rejection of cells, but a productization strategy: shifting the therapeutic unit from a living, variable entity to a more controllable biologic—often EVs—while retaining the key MSC mechanisms (immunomodulation, reparative signaling). This transition is being shaped by reproducibility standards, the rise of scalable purification platforms, and increasingly explicit regulatory expectations for potency, comparability, and risk-based development [5].

1.1 Stem Cell Classes and Comparative Attributes

1.1.1 Conceptual Taxonomy

Modern classification can be organized by potency (capacity to generate differentiated lineages), developmental origin, and manufacturing/clinical risk profile (Table 1.1). Potency is commonly described as (i) totipotent (embryo + extraembryonic lineages), (ii) pluripotent (three germ layers), and (iii) multipotent/oligopotent/unipotent (restricted lineage range) [6]. In translational terms, two distinctions repeatedly drive clinical strategy (Fig. 1.1):

1. Pluripotent platforms (ESCs, iPSCs) primarily enable cell replacement (or tissue reconstruction) because they can generate large, standardized batches of specialized cell types, but they bring higher requirements for genomic stability, purity, tumorigenicity risk management, and long-term follow-up [2].
2. Adult stem/progenitor platforms (HSCs, NSCs, MSCs) typically emphasize repair orchestration (immune regulation, trophic support, niche repair) and are often deployed under different regulatory and manufacturing paradigms, especially when minimally manipulated and used homologously [7].

1.2 Why MSCs Became the Leading Adult Stem Cell Platform

1.2.1 Platform Dominance as a Convergence of Biology and Manufacturability

MSCs became the dominant adult stem cell platform not because they are the most "stem-like" adult cells, but because their therapeutic value often does not require durable engraftment. The ISCT committee framing explicitly distinguishes rare "mesenchymal stem" cells (self-renewal + differentiation) from bulk

Table 1.1 Comparative overview of major stem cell classes: potency, sources, defining markers, therapeutic logic, clinical applications, and key limitations

Stem cell class	Typical potency	Common sources	Defining markers (examples used in practice)	Dominant therapeutic logic	Clinical status and examples	Key limitation/risks
Embryonic stem cells (ESCs)	Pluripotent	Pre-implantation embryo (blastocyst inner cell mass)	Core pluripotency network (OCT3/4, SOX2, NANOG) and surface markers (TRA-1-60/TRA-1-81, SSEA3/4) used in PSC characterization	Differentiation into specialized cells for replacement	hPSC trials are increasing; major trial focus includes the eye and CNS [2]	Ethical constraints; immune rejection unless autologous/immune evasive; residual undifferentiated cells → tumorigenicity risk; stringent quality controls needed
Induced pluripotent stem cells (iPSCs)	Pluripotent	Reprogrammed adult somatic cells	The same PSC marker logic (OCT4, SOX2, NANOG; TRA-1-60; SSEA4) used for validation	Patient-specific or banked allogeneic PSC sources for standardized differentiation	Rapid expansion of iPSC clinical translation across multiple therapeutic areas; trials summarized with registry identifiers in recent reviews [8]	Reprogramming-associated genomic/epigenomic variation; manufacturing complexity; tumorigenicity management; cost and time for autologous approaches
Hematopoietic stem cells (HSCs)	Multipotent (blood lineages)	Bone marrow, mobilized peripheral blood, umbilical cord blood	Clinical/experimental enrichment often uses CD34+ with additional phenotyping (context dependent)	Curative hematologic reconstitution (transplant) and, increasingly, platforms for gene-edited autologous therapies	HSCT is a potentially curative modality for many hematologic diseases and malignancies; expanding role in immune-mediated disease contexts [9]	Graft-versus-host disease (allogeneic); conditioning toxicity; infection risk; relapse; logistics and donor availability

(continued)

Table 1.1 (continued)

Stem cell class	Typical potency	Common sources	Defining markers (examples used in practice)	Dominant therapeutic logic	Clinical status and examples	Key limitation/risks
Neural stem/progenitor cells (NSCs)	Multipotent (neural lineages)	Fetal tissue–derived sources; PSC-derived neural progenitors; adult brain niches	Frequently reported markers include SOX2 and nestin in NSC biology; marker interpretation is context and state dependent in single-cell studies	Neural repair via cell replacement and/or trophic support	Clinical development remains challenging; PSC-derived dopaminergic progenitors/neurons are advancing for Parkinson's disease [10]	Cell delivery constraints; integration and circuit-level safety; tumorigenicity risk with PSC-derived inputs; need for long-term follow-up
Mesenchymal stromal/stem cells (MSCs)	Multipotent (limited mesenchymal lineages in vitro); clinically often "stromal/signaling" logic	Bone marrow, adipose tissue, perinatal tissues	ISCT-associated minimal criteria (plastic adherence; CD73/CD90/CD105 positive; absence of CD11b/CD14/CD19/CD34/CD45/CD79a/HLA-DR; trilineage differentiation in vitro)	Immunomodulation and paracrine repair orchestration; "medicinal signaling" effects in vivo	Broad global clinical testing; first US MSC approval: Ryoncil (remestemcel-L-rknd) for pediatric SR-aGVHD [4]	Heterogeneity (donor/tissue/culture); variable potency; pulmonary first-pass effect after IV delivery; procoagulant risk (TF/CD142); limited durable engraftment

Abbreviations: *ESCs* embryonic stem cells, *iPSCs* induced pluripotent stem cells, *HSCs* hematopoietic stem cells, *NSCs* neural stem/progenitor cells, *MSCs* mesenchymal stromal/stem cells, *PSCs* pluripotent stem cells, *hPSC* human pluripotent stem cells, *HSCT* hematopoietic stem cell transplantation, *SR-aGVHD* steroid-refractory acute graft-versus-host disease, *ISCT* International Society for Cell & Gene Therapy, *CNS* central nervous system, *TF/CD142* tissue factor (cluster of differentiation 142), *HLA-DR* human leukocyte antigen-DR isotype, *SSEA* stage-specific embryonic antigen, *OCT3/4 (OCT4)* octamer-binding transcription factor 4, *SOX2* SRY-box transcription factor 2; *NANOG* homeobox protein NANOG

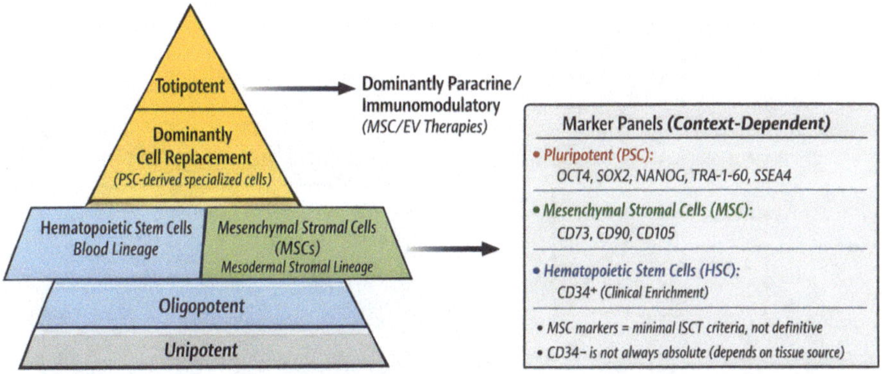

Fig. 1.1 Refined potency hierarchy and translational "product logic" across stem cell platforms. Schematic representation of stem cell potency and corresponding therapeutic "product logic." The hierarchy spans from pluripotent to oligopotent and unipotent states. The totipotent stage is shown for conceptual orientation, whereas the translational emphasis is placed on pluripotent and adult stem/progenitor platforms. Pluripotent stem cells (PSCs) are positioned at the apex and are primarily associated with cell replacement strategies via differentiation into specialized cell types, although paracrine effects may also contribute. In contrast, hematopoietic stem cells (HSCs) and mesenchymal stromal/stem cells (MSCs) are depicted as parallel multipotent lineages rather than hierarchically ordered, reflecting their distinct developmental origins (hematopoietic vs mesodermal stromal). The MSC-based therapies are shown as predominantly paracrine and immunomodulatory, including EV-mediated mechanisms, while retaining limited differentiation capacity. An inset panel summarizes commonly used context-dependent marker sets: PSCs (OCT4, SOX2, NANOG, TRA-1-60, SSEA4), MSCs (CD73, CD90, CD105, based on minimal International Society for Cell & Gene Therapy (ISCT) criteria), and HSCs (CD34+ as a clinical enrichment marker). It is emphasized that these markers are not definitive and may vary by tissue source, culture conditions, and analytical methods; for example, CD34 negativity in MSCs is not absolute

"mesenchymal stromal" populations whose clinical relevance is largely secretory, immunomodulatory, and homing related [7]. This positioning redefined MSCs as a biologic response modulator compatible with allogeneic "off-the-shelf" manufacturing and repeat dosing—features that map well onto clinical needs in inflammatory, ischemic, and immune-mediated diseases [1].

1.2.2 Mechanistic Pillars

Immunomodulation as Context-Dependent "Licensing"

A consistent theme is that MSC immunomodulation depends on inflammatory context ("licensing/priming"), with cytokines such as IFN-γ/TNF-α upregulating pathways that correlate with suppression of immune effector function. Reviews synthesize multi-cell targeting (T cells, monocytes/macrophages, dendritic cells) and soluble mediators as central mechanisms [11] (Fig. 1.2). Primary work further suggests that potency-relevant readouts can be linked to specific molecules: for

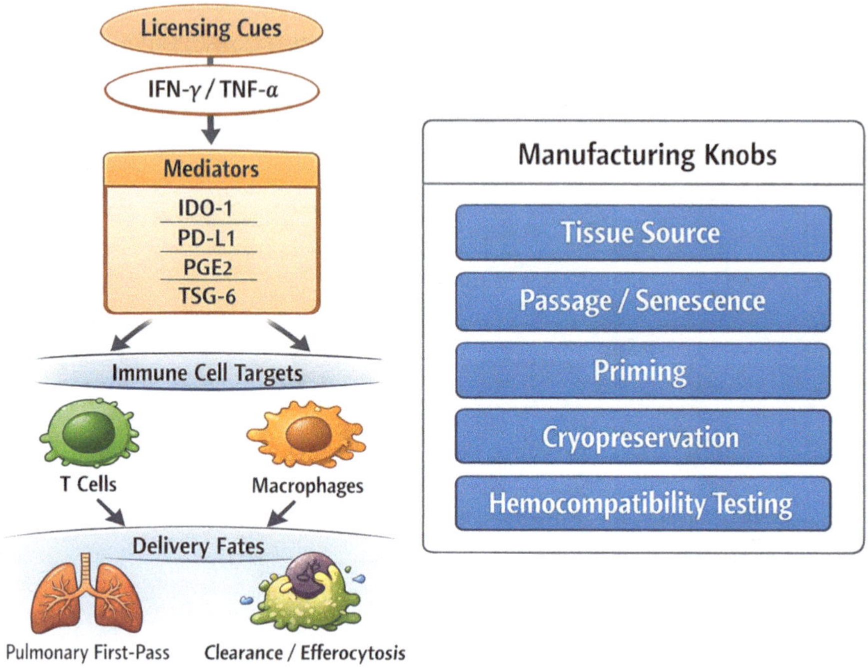

Fig. 1.2 Mechanisms of mesenchymal stromal/stem cell (MSC) therapeutic action and key controllable variables. This schematic illustrates the integrated mechanisms underlying MSC therapeutic effects. Pro-inflammatory licensing cues, primarily Interferon γ (IFN-γ) and Tumor Necrosis Factor α (TNF-α), activate MSCs to produce key immunomodulatory mediators including Indoleamine 2,3-dioxygenase-1 (IDO-1), Programmed Death-Ligand 1 (PD-L1), Prostaglandin E2 (PGE2), and Tumor Necrosis Factor–Stimulated Gene 6 (TSG-6). These mediators modulate immune cell targets such as T cells and macrophages, promoting immunosuppression and resolution of inflammation. Following systemic administration, MSCs undergo characteristic delivery fates, including pulmonary first-pass entrapment and subsequent clearance through efferocytosis. The right panel highlights critical "manufacturing knobs" that influence MSC potency and clinical performance, including tissue source, passage number and cellular senescence, priming/preconditioning strategies, cryopreservation methods, and hemocompatibility testing. Together, these factors define the therapeutic efficacy and reproducibility of MSC-based interventions

example, IFN-γ-licensed MSC expression of IDO-1 and PD-L1 correlated with suppression of third-party T-cell proliferation in a potency-oriented analysis [12]. Particularly for intravascular administration, MSC-associated tissue factor (CD142) expression can trigger coagulation cascades and instant blood-mediated inflammatory reaction (IBMIR), representing a clinically relevant safety concern requiring standardized hemocompatibility testing.

Paracrine Effects and "Hit-and-Run" Biology

Intravenously administered MSCs often display limited residence time, with many cells trapped in lung vasculature and cleared within ~24 to 48 h in animal models; nevertheless, immune reprogramming can follow clearance, supporting a "hit-and-run" paradigm in which short-lived MSCs trigger longer-lived host responses [13]. An increasingly discussed mechanism is that apoptotic MSCs and their clearance by phagocytes (efferocytosis) can be immunologically active, shifting emphasis from "live cell engraftment" to "host re-education" [14].

Homing and Biodistribution as Both Opportunity and Constraint
Classical narratives emphasized injury-site homing; newer biodistribution work stresses that IV delivery commonly produces a pulmonary first-pass effect, limiting systemic homing yet potentially enabling lung–immune interactions that propagate systemic effects [13].

Safety Profile and the Evolution of Risk Thinking
Compared with pluripotent platforms, MSCs are generally viewed as lower tumorigenicity risk; however, safety concerns are not absent and increasingly focus on (i) infusion/hypersensitivity reactions, (ii) infectious agent transmission (donor derived, process related), (iii) ectopic tissue formation, and (iv) hemocompatibility/thrombotic risk. The FDA's Summary Basis for Regulatory Action for Ryoncil lists "hypersensitivity/acute infusion reactions," "transmission of infectious agents," and "ectopic tissue formation" as major risks, illustrating how regulators translate theoretical hazards into labeled safety constructs [15]. Separately, mechanistic reviews detail MSC expression/release of tissue factor and interactions with coagulation pathways, motivating product characterization for hemocompatibility—particularly for intravascular dosing [16].

1.2.3 Manufacturing and Scalability as Decisive Advantages

MSCs align unusually well with modern biomanufacturing because many programs can use allogeneic donors, banked master/working cell banks, and scalable expansion systems. Recent manufacturing literature emphasizes the need to (1) map process parameters to product critical quality attributes (CQAs), (ii) manage senescence and culture adaptation, and (iii) link in vitro potency assays to clinical mechanism and indication [17]. The ISCT nomenclature paper also pushes toward tissue source annotation and robust functional assay matrices informed by intended mechanism of action, reflecting an industry shift from surface marker minimalism to mechanism-linked release testing [7].

1.2.4 Clinical Translation Signals and a Regulatory Inflection Point

Ryoncil: First US MSC Approval (Pediatric SR-aGVHD)
On December 18, 2024, the FDA approved Ryoncil (remestemcel-L-rknd) for pediatric SR-aGVHD, explicitly identifying it as an allogeneic bone marrow–derived MSC therapy [4]. The FDA's regulatory action summary describes reliance on a single-arm study (MSB-GVHD001) with an overall response rate endpoint at day 28 and notes postmarketing requirements/commitments including assay development—highlighting how potency and assay adequacy remain active regulatory topics even at approval [15].

A Cautionary Counterexample: Alofisel (Darvadstrocel) Withdrawal in the EU
Europe's earlier MSC-based product darvadstrocel (Alofisel) illustrates that authorization does not end evidentiary uncertainty; the European Medicines Agency [18] reported withdrawal from the EU market in December 2024 [19]. Company communications attribute the withdrawal to confirmatory trial outcomes not meeting primary endpoints, underscoring how heterogeneous indications, endpoints, and product comparability can destabilize the benefit–risk profile over time [20].

1.3 Emergence of EVs as Therapeutic Mediators

1.3.1 EV Definition, Biogenesis, and the Nomenclature Discipline

EVs are lipid bilayer–delimited particles released by cells; however, assigning biogenesis pathways ("exosome" vs "microvesicle/ectosome") typically requires evidence that is difficult to obtain in most therapeutic development contexts. MISEV2018 therefore recommends EV as a generic term and urges operational subtype definitions based on size, density, biochemical markers, or source conditions rather than assuming origin-specific labels [5]. This recommendation has direct translational consequences: if a therapeutic developer cannot convincingly specify biogenesis, the regulatory and scientific focus shifts from naming purity to measuring composition, contaminants, and function in a reproducible way [5]. MISEV2023, developed through broad international community input, expands guidance across EV production, separation, characterization, EV release/uptake, and functional studies, reflecting the field's movement toward a quality system mindset similar to biologic development [21, 22].

1.3.2 Cargo and Mechanisms of Action

Therapeutic EV interest is driven by their role as carriers of proteins, lipids, and nucleic acids that can reprogram target cells, with attractive features such as biocompatibility and low mutagenicity risk emphasized in recent quality/safety analyses [23]. In MSC biology specifically, EVs are frequently treated as mediators of the MSC paracrine hypothesis, potentially transmitting immunoregulatory signals to immune cells via surface ligands and internal cargo [24]. A key analytical point for translational readers is that "EV mechanism" is not a single mechanism: depending on product and indication, EV effects may be mediated by (i) direct receptor–ligand signaling at the EV surface, (ii) uptake and cytosolic delivery of functional components, or (iii) modulation of extracellular proteostasis and coagulation/inflammation pathways through co-isolated non-vesicular components—precisely why MISEV emphasizes contaminant control and functional attribution [5].

1.3.3 Advantages and Trade-Offs Versus Cell Therapy

Why EVs Are Attractive as Cell-Free Therapy EVs cannot replicate and are often framed as lower risk for uncontrolled proliferation compared with living cells. They may be easier to store and distribute, and they can be engineered or formulated as drug delivery systems [23]. From a development perspective, EVs potentially reduce risks tied to live cell biodistribution while preserving desired signaling functions, supporting the strategic move from "cell as drug" to "vesicle as drug" [13].

What EVs Do Not Automatically Solve EV products are heterogeneous, and many separation methods co-isolate non-vesicular particles and soluble proteins. MISEV2018 stresses the need to demonstrate enrichment of EV markers and depletion of contaminants when claiming function [5]. Thus, the technical bottleneck shifts from "keeping cells alive" to "defining, quantifying, and releasing a complex biologic nanoparticle mixture with validated potency" [5].

However, EVs are not inherently risk-free; their cargo may include bioactive molecules capable of modulating oncogenic, inflammatory, or coagulation pathways, necessitating rigorous safety characterization.

1.3.4 Isolation, Characterization, and Potency Assays

Isolation and Enrichment Methods
MISEV2018 discusses a wide range of separation approaches (differential ultracentrifugation, density gradients, filtration, size exclusion chromatography, immunoisolation), emphasizing that no one-size-fits-all method exists and that combinations

may optimize recovery vs specificity depending on the clinical context [5]. For therapeutic-scale manufacturing, recent translational work increasingly prioritizes scalable unit operations such as tangential flow filtration, size exclusion chromatography, and chromatography-based purification platforms for robust EV production.

Characterization Expectations
MISEV2018 recommends multi-technique characterization—quantifying particles, assessing protein/lipid content, measuring EV-associated markers, and evaluating non-vesicular components. It also highlights "topology" testing to determine whether putative active components are inside vesicles or surface exposed, because topology affects functional interpretation and assay design [5].

Potency Assays (EVs)
EV potency assays remain indication and mechanism specific, but the regulatory trajectory is clear: potency should be assured by a science- and risk-based strategy incorporating process controls, in-process tests, and lot release assays (Fig. 1.3). The FDA's 2023 draft guidance formalizes potency assurance as a multifaceted strategy designed to ensure each lot has the "specific ability or capacity" to achieve the intended therapeutic effect [25]. For EV developers, this implies (i) anchoring

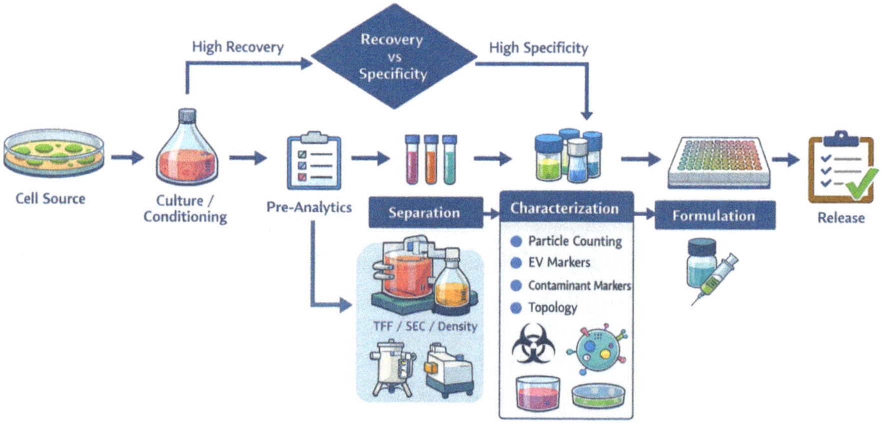

Fig. 1.3 EV production-to-release workflow aligned with MISEV guidelines and potency assurance framework. This schematic illustrates the standardized workflow for extracellular vesicle (EV) production, processing, and release in accordance with the Minimal Information for Studies of Extracellular Vesicles (MISEV) recommendations and potency assurance principles. The process begins with the selection of an appropriate cell source, followed by culture and conditioning to optimize EV secretion. Pre-analytical variables are controlled prior to EV isolation using separation techniques such as Tangential Flow Filtration (TFF), Size Exclusion Chromatography (SEC), or density-based methods. A critical decision node highlights the trade-off between recovery and specificity during isolation. Subsequent characterization includes particle quantification, assessment of EV-associated markers, evaluation of contaminant markers, and analysis of vesicle topology. Formulation strategies prepare EVs for downstream application, followed by a potency assay matrix designed to confirm biological activity and consistency. Final product release is contingent upon meeting predefined quality, safety, and functional criteria. Notably, classification of vesicles as "EVs" or "microvesicles" requires direct evidence of their biogenesis pathways

assays to a defined mechanism or clinically relevant activity, (ii) building reference materials and system suitability controls, and (iii) bridging analytical attributes to functional readouts—often requiring an "assay matrix" rather than a single test [25].

1.4 Transition from Bench to Clinical Translation

1.4.1 Preclinical Models as Translational Instruments Rather than Proof-of-Concept Demonstrations

High-quality translation treats preclinical models as risk reduction tools: clarifying mechanism, biodistribution, dose–response, and safety signals that can inform clinical monitoring and release tests. For MSCs, biodistribution findings (pulmonary first-pass, rapid clearance) directly impact clinical expectations regarding homing, dosing frequency, and the plausibility of replacement vs signaling mechanisms [13]. For EVs, preclinical inference must also confront functional attribution: if co-isolated contaminants could explain effects, then both mechanistic claims and potency assays become unstable—hence the need for MISEV-aligned controls [5].

1.4.2 GMP Manufacturing and Quality Systems

United States: Potency and Comparability as Core CGT Constraints

Recent FDA draft guidance defines potency assurance as a broad strategy spanning process design, control, in-process testing, and lot release assays, aiming to ensure each lot's intended therapeutic capacity [25]. In parallel, the FDA's 2023 draft guidance on manufacturing changes and comparability emphasizes that CGT products face unique comparability challenges because of limited attribute knowledge, variable starting materials, limited shelf life, and manufacturing complexity. It stresses prospective risk assessment and comparability studies, noting that analytical data alone may be insufficient and that nonclinical or clinical bridging may be required [26]. A practical implication for MSC/EV programs is that scaling (new bioreactor, new medium, new separation column, new cryoprotectant, new fill–finish site) is not "process improvement" unless comparability is demonstrated; otherwise, it risks becoming a "new product" scenario requiring new development or regulatory pathways [26].

Regulatory Boundary Conditions: HCT/P Enforcement and "Minimal Manipulation"

The FDA's HCT/P guidance clarifies minimal manipulation (structural vs cellular/nonstructural tissue definitions) and homologous use and documents enforcement discretion extensions, reflecting active regulatory efforts to separate legitimate product development from high-risk unapproved interventions [3]. The same surgical procedure exception guidance defines a narrow exception requiring removal and implantation into the same individual during the same surgical procedure with limited processing, clarifying what is and is not exempt from Part 1271 requirements [27].

Expedited Pathways: RMAT and the Post-cures Act Environment
The twenty-first Century Cures Act was signed into law on December 13, 2016, and authorized programs such as RMAT designation [28]. The FDA's RMAT description outlines eligibility criteria (regenerative therapy definition, serious disease, preliminary clinical evidence of unmet need potential) and ties RMAT requests to IND submission timing—showing how US policy attempts to accelerate legitimate regenerative products while maintaining evidence thresholds [29]. The FDA's regenerative medicine framework page consolidates the agency's policy framework and cites final guidance on minimal manipulation/homologous use, the same surgical procedure exception, and expedited programs [30].

European Union: Risk-Based GMP and Investigational ATMP Development
The European Commission's GMP guideline specific to ATMPs (adopted November 22, 2017, operational May 22, 2018) frames ATMP manufacturing within a tailored GMP system, explicitly noting that these guidelines are specific to ATMPs and emphasizing GMP requirements for investigational products and "hospital exemption" manufacturing under equivalent quality standards [31]. The EMA guideline on investigational ATMPs in clinical trials (adopted January 20, 2025; effective July 1, 2025) emphasizes a multidisciplinary structure covering quality, nonclinical, and clinical documentation and promotes a risk-based approach wherein the extent of data can be adapted to identified risks, with priority on subject safety [32].

1.4.3 Timeline of Translational Milestones 2016–2026

Key milestones below highlight policy, standards, and clinical proof points that shaped the cell-to-EV transition (Fig. 1.4). Sources cover US law and FDA guidance actions, EU GMP/ATMP guidance, EV standards (MISEV), and landmark clinical publications/approvals (Fig. 1.5).

1.5 Clinical Landscape and Evidence Themes

1.5.1 PSCs: Expansion with Cautious Optimism

The most data-rich high-level signal in recent years is the expansion of hPSC clinical testing: a 2025 update reports 115 regulatory authority–approved trials testing 83 hPSC products, with the majority targeting the eye, the central nervous system,

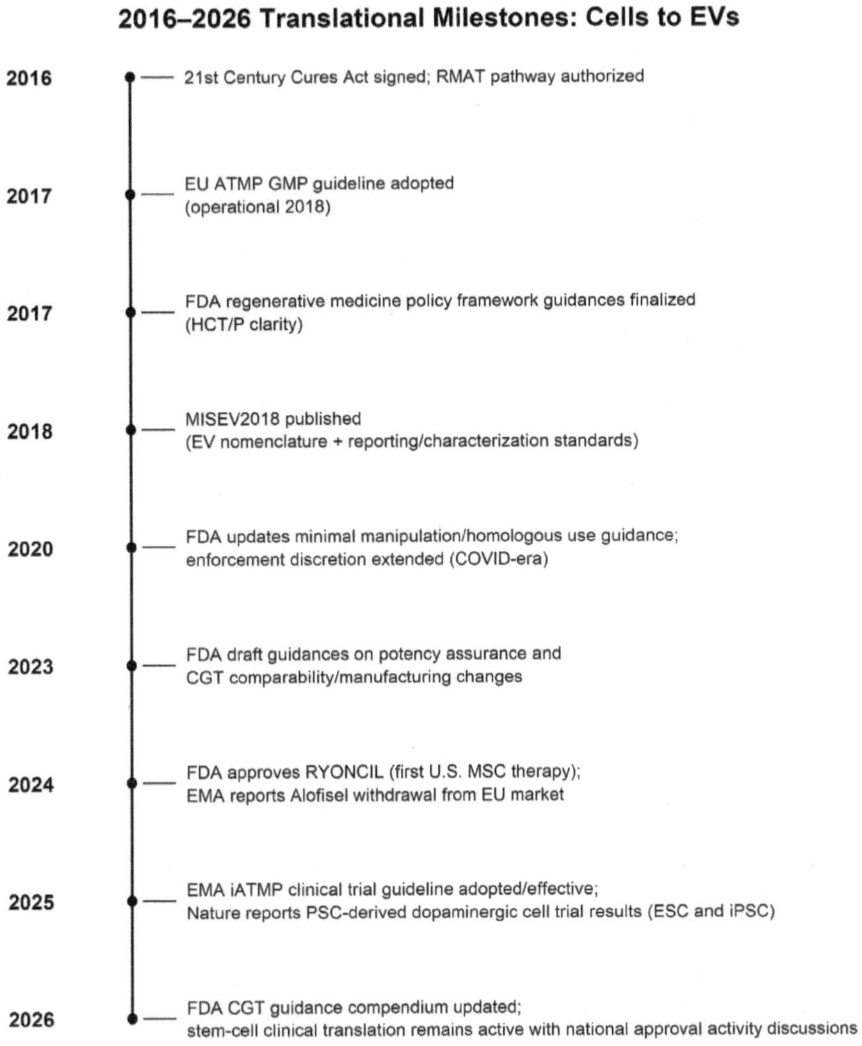

Fig. 1.4 The 2016–2026 translational milestones in regenerative medicine: from cell therapies to extracellular vesicles. Vertical timeline summarizing key regulatory and scientific milestones shaping regenerative medicine between 2016 and 2026. Highlights include major policy frameworks (FDA, EMA), standardization efforts such as MISEV2018, evolving guidance on manufacturing and potency, and landmark clinical developments in mesenchymal stromal/stem cell (MSC) and pluripotent stem cell (PSC) therapies, reflecting the field's progression toward extracellular vesicle (EV)–based therapeutics

and cancer and with >1200 patients dosed and no generalizable safety concerns to date (while still requiring product-specific risk management and long-term follow-up) [2]. Neurology has emerged as a prominent testbed for PSC-derived cell replacement. In 2025, a first-in-human report described an allogeneic iPSC-derived

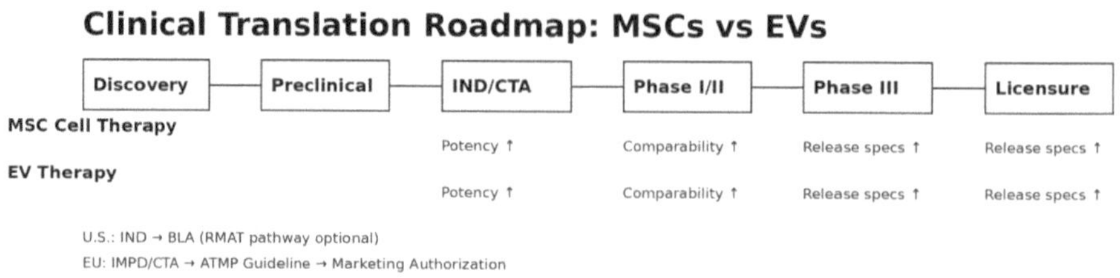

Fig. 1.5 Clinical translation roadmap for MSC and extracellular vesicle therapies: from discovery to licensure. Stage-gated schematic illustrating the clinical translation pathway for mesenchymal stromal/stem cell (MSC) therapies and extracellular vesicle (EV)–based therapeutics, spanning discovery, preclinical development, IND/CTA submission, Phase I/II, Phase III, and licensure. Parallel tracks highlight shared and distinct development considerations, including the progressive intensification of comparability requirements, potency assay development, and release specifications. Key regulatory touchpoints are indicated for the United States (IND/BLA pathways, including RMAT designation) and the European Union (IMPD/CTA processes and ATMP regulatory framework), reflecting harmonized yet region-specific translational strategies

dopaminergic progenitor approach showing survival, dopamine production, and no tumor formation in a phase I/II setting [10]. A translational inference is that PSC programs increasingly resemble advanced biologic programs: success depends on differentiation fidelity, elimination of undifferentiated cells, immune management, and long-term surveillance—moving the field toward more standardized product characterization frameworks [2].

1.5.2 MSCs: Real Approvals amid Variable Efficacy Narratives

MSCs occupy a different evidentiary niche: many trials across diverse indications with heterogeneity in cell source, dosing, delivery route, and endpoints, often producing mixed efficacy outcomes while maintaining a relatively consistent safety narrative. Recent clinical/regenerative reviews continue to emphasize MSC versatility and safety considerations (including tissue sourcing diversity and immunomodulatory functions) as drivers of continued investment [33]. The FDA approval of Ryoncil provides the clearest signal that MSC products can reach modern regulatory standards when the indication, endpoint strategy, and CMC controls are aligned. The FDA action summary describes reliance on a defined clinical endpoint (day 28 overall response rate) and includes CMC-related postmarketing requirements and assay development commitments—again showing potency as a living regulatory issue [15]. At the same time, the EU withdrawal of Alofisel highlights a countervailing theme: confirmatory evidence requirements and endpoint selection can overturn earlier market access, emphasizing the need for robust trial design, patient stratification, and product consistency [19, 20].

1.5.3 *EVs: Early Clinical Experimentation and Registry-Level Visibility*

Clinical translation of EV therapies remains early-phase but is now visible in official registries, including trials evaluating the following approaches (Fig. 1.6):

- MSC-derived EV products administered intravenously for acute respiratory distress–related conditions [34]
- Nebulized/inhaled MSC-derived EV/exosome preparations for post-COVID and respiratory indications [35]

These entries do not, by themselves, validate efficacy, but they are a strong signal that developers consider EVs sufficiently "productizable" to support early human dosing, raising immediate questions about manufacturing consistency, identity/impurity controls, and potency assay selection [5].

A recent bench-to-bedside mapping of the MSC–EV field by Mussin et al. [36] illustrates the transition from broad preclinical efficacy studies toward defined therapeutic products, scalable manufacturing, mechanism-linked potency testing, and early human evaluation. The analysis emphasizes that clinical advancement will depend on harmonized nomenclature, transparent reporting of production and purification methods, standardized dose metrics, and clinical designs capable of linking product attributes with exposure and patient outcomes.

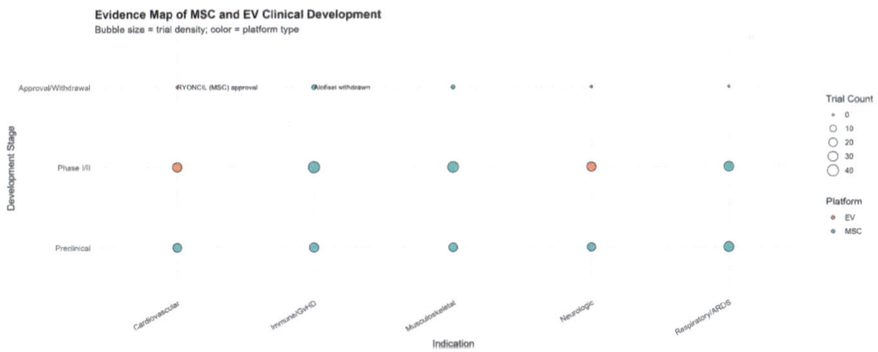

Fig. 1.6 Evidence map of mesenchymal stromal/stem cell (MSC) and extracellular vesicle (EV) clinical development across indications and maturity stages. Bubble plot illustrating the distribution of clinical evidence across major therapeutic indications (immune/GVHD, musculoskeletal, respiratory/ARDS, cardiovascular, and neurologic) and development stages (preclinical, phase I/II, and approval/withdrawal). Bubble size represents trial density, while color denotes therapeutic platform (MSC vs EV vs PSC derived). Key regulatory anchors include the approval of Ryoncil and the market withdrawal of Alofisel, highlighting the dynamic and evolving translational landscape

1.6 Challenges and Future Directions

1.6.1 Mechanism, Heterogeneity, and the Potency "Translation Gap"

Both MSC and EV programs repeatedly encounter a potency translation gap: (i) biological activity is multifactorial, (ii) assays can be poorly predictive across indications, and (iii) process changes can shift product composition in ways that are difficult to detect with limited analytics. The FDA potency assurance draft explicitly positions potency as a lifecycle strategy (process design + controls + assays), implying that "one release assay" is rarely adequate for complex CGT products [25]. The FDA comparability draft adds that CGT products face unique barriers (variable starting materials, limited shelf life, limited attribute knowledge), warning that analytical comparability alone may be insufficient in some cases—forcing nonclinical/clinical bridging [26]. For MSCs, heterogeneity sources include tissue origin, donor biology, senescence, culture conditions (2D vs 3D), and priming; manufacturing reviews increasingly frame these as controllable "process levers" that must be tied to CQAs and indication-specific mechanisms [37, 38]. For EVs, heterogeneity expands to separation method artifacts and co-isolated contaminants—hence the MISEV-driven insistence on marker/contaminant panels and topology/function attribution [5].

1.6.2 Delivery Biology as a Design Constraint

The pulmonary first-pass effect for IV MSC delivery is not just a mechanistic curiosity; it affects dose selection, safety monitoring, and the rationale for targeting distal organs. If MSCs act through immune reprogramming following lung entrapment/clearance, then endpoints and biomarkers should reflect immune dynamics rather than expecting cell engraftment at injury sites [13]. Analogously, EV biodistribution and uptake mechanisms must be addressed explicitly, particularly for inhaled vs IV products, where local exposure and systemic spillover require different safety strategies [23].

1.6.3 Regulatory Convergence toward "Quality System Thinking"

Across regions, guidance is converging toward risk-based expectations: the EMA investigational ATMP guideline emphasizes risk-based adaptation of data requirements and multidisciplinary documentation across quality, nonclinical, and clinical domains [32]. The EU ATMP GMP guidelines highlight that GMP is mandatory for authorized and investigational medicinal products and that ATMPs manufactured under hospital exemption must meet equivalent quality standards, reinforcing that "local" or "academic" development does not exempt products from quality rigor [18]. In the United States, the FDA's regenerative medicine framework situates HCT/P enforcement clarity alongside expedited development tools, indicating a dual strategy: facilitate legitimate translation while restricting unapproved, high-risk interventions [30].

1.6.4 Outlook: From Cell-Free Therapy to Programmable Biologics

The field appears to be moving toward programmable biologics rather than simply "cells vs EVs." Likely trajectories include:

- Engineered MSCs and EVs with tuned immunomodulatory functions, provided comparability and potency constraints can be met [25, 26].
- Increasing use of platform manufacturing (standardized upstream cell factories, scalable downstream purification) to reduce batch variability for EV therapeutics [23].
- For PSC-derived products, expanding trial maturity in CNS disorders and other high-need areas should be paired with explicit long-term follow-up and surveillance models [10].

The central message for translational readers is that "regenerative medicine" is increasingly less about differentiation potential and more about controllable, testable mechanisms of tissue repair—a shift that naturally favors platforms (like MSC-derived EVs) that can be standardized, potency tested, and scaled under evolving CGT regulatory frameworks [5, 7].

Chapter 2
MSCs: Biology, Sources, and Functional Properties

Abstract This chapter examines the biological characteristics, sources, and functional properties of mesenchymal stromal/stem cells (MSCs). It discusses the operational definition of MSCs, ISCT criteria, and emerging perspectives on nomenclature and heterogeneity. Comparative analysis of tissue sources highlights differences in proliferation, immunomodulation, and secretome composition. The chapter also explores differentiation potential, homing behavior, and immunoregulatory mechanisms, emphasizing the predominance of paracrine effects. Advantages, limitations, and clinical implications of MSC-based therapies are critically evaluated.

Keywords MSC biology · Stem cell sources · Immunomodulation · Differentiation · Regenerative medicine

Mesenchymal stem/stromal cells (MSCs) are operationally defined cell products that are typically isolated from diverse tissues, expanded ex vivo by adherence to plastic, and released as "MSC" based on a narrow set of phenotypic criteria plus in vitro differentiation assays. Contemporary work strongly supports two overarching conclusions: (i) MSCs are not a single, uniform biological entity but a heterogeneous ensemble whose composition depends on tissue source, donor factors, and manufacturing conditions and (ii) the most established in vivo therapeutic effects are mediated predominantly through context-dependent paracrine and immunomodulatory mechanisms rather than durable engraftment and cell replacement [7].

The most widely used identity framework remains the minimal criteria articulated by the International Society for Cell & Gene Therapy (ISCT)—plastic adherence under standard culture conditions; a "canonical" immunophenotype (CD73/CD90/CD105 positive; hematopoietic/endothelial markers negative); and trilineage differentiation into osteogenic, chondrogenic, and adipogenic lineages in vitro. However, the last decade of literature demonstrates that these criteria are insufficient to predict potency, mechanism of action (MoA), or clinical effectiveness, motivating multiple efforts toward (a) stronger nomenclature discipline ("stromal" vs "stem"), (b) tissue source annotation, (c) functional assay matrices aligned to intended MoA, and (d) improved peer-reviewed reporting standards and consensus definitions for clinical trials [7].

N. M. Mussin, A. Tamadon, *Mesenchymal Stem Cells and Their Extracellular Vesicles*, https://doi.org/10.1007/978-3-032-36437-1_2

Comparisons across sources reveal consistent, source-dependent trends. Adult tissue MSCs—especially bone marrow–derived MSCs (BM–MSCs) and adipose-derived MSCs (AT/AD–MSCs)—dominate historical development and many clinical programs. Perinatal sources (umbilical cord/Wharton's jelly and placenta-derived MSCs) often show higher proliferative capacity and expansion efficiency, as well as "younger" phenotypes, but are not interchangeable with adult MSCs in immunophenotype and secretome composition. Proteomic studies show that fetal/perinatal MSC secretomes can be more diverse, and source-specific secretome "signatures" are repeatedly observed even when cells meet ISCT identity thresholds [39].

Mechanistic studies in the last decade refine how MSCs work in vivo: systemic infusion frequently leads to early lung sequestration and rapid clearance, shifting attention toward short-lived but potent immunoregulatory cascades—including apoptosis/efferocytosis-dependent effects, macrophage polarization, and broad paracrine signaling through soluble mediators and EVs. Licensing/priming by inflammatory cytokines can enhance immunomodulatory factors yet can also increase immunogenic markers depending on culture supplements and protocols, underscoring the need for structured critical quality attributes (CQAs) and potency assays [40].

Clinically, MSC products have generally shown favorable short-term safety, particularly for intravascular administration, across broad indications, while efficacy signals remain inconsistent and indication dependent. Recent late-stage "setbacks" (including a negative phase III trial in Crohn's complex perianal fistulas) and regulatory dynamics have intensified calls for better potency-linked release testing and standardized reporting. Meanwhile, a landmark regulatory milestone occurred when the US Food and Drug Administration [41] approved Ryoncil (remestemcel-L-rknd) in December 2024 for pediatric steroid-refractory acute GVHD, reflecting an evolving (but still stringent) pathway for MSC-based biologics [42].

2.1 Definition and Minimal Criteria for MSCs

2.1.1 Operational Definition and Nomenclature Discipline

In modern research and clinical translation, "MSC" most often denotes a manufactured, culture-expanded stromal cell product rather than a single, uniquely defined in vivo stem cell entity. This distinction is central to nomenclature debates and the push to preserve "MSC" as an acronym while clarifying whether it means "mesenchymal stromal/stem cell" depending on demonstrated differentiation potential at appropriate rigor [7].

A widely cited reframing argues that many clinical MSC preparations function primarily as "medicinal signaling cells," emphasizing trophic and immunomodulatory roles over direct tissue replacement [43]. Contemporary critiques of this renaming emphasize that "medicinal signaling" captures an important MoA but may overcompress the breadth of MSC biology and context specificity; the key translational point remains that paracrine/immune mechanisms dominate many therapeutic scenarios [44].

2.1.2 ISCT Minimal Criteria and What They Do—And Do Not—Guarantee

The ISCT minimal criteria specify that MSCs (Fig. 2.1) (1) adhere to plastic under standard culture conditions; (2) express canonical "positive" markers CD73, CD90, and CD105 and lack hematopoietic/endothelial markers (commonly CD45, CD34, CD14/CD11b, and CD19/CD79a and typically HLA-DR as a negative/release criterion); and (3) demonstrate in vitro differentiation into osteogenic, chondrogenic, and adipogenic lineages [7].

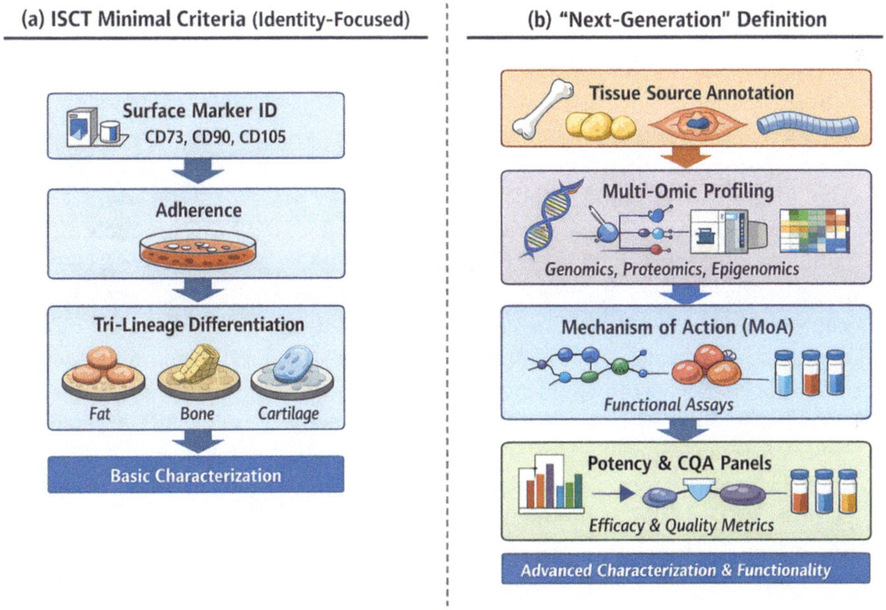

Fig. 2.1 Comparative framework for mesenchymal stromal/stem cell (MSC) definition: from ISCT minimal criteria to next-generation multi-omic and functional characterization. This schematic contrasts (**a**) the International Society for Cell & Gene Therapy (ISCT) minimal criteria—primarily focused on MSC identity through plastic adherence, surface marker expression (CD73, CD90, CD105), and trilineage differentiation—with (**b**) emerging next-generation definition frameworks. The advanced approach integrates tissue source annotation, comprehensive multi-omic profiling (including genomics, proteomics, and epigenomics), and mechanism-of-action (MoA)–linked functional assays. These are further connected to potency testing and critical quality attribute (CQA) panels, enabling a more robust, reproducible, and clinically predictive characterization of MSC products

However, major limitations are now well characterized. First, the canonical immunophenotype is not specific to a single functional state and does not uniquely separate MSCs from fibroblast-like stromal populations or predict potency across disease contexts. This is one motivation for expanding marker panels and integrating multi-omic characterization in both research and GMP settings [45]. Second, the "HLA-DR negative" convention can be protocol dependent. For example, replacement of fetal bovine serum with human platelet lysate (a common xeno-free expansion strategy) can increase the fraction of MSCs expressing MHC class II/ HLA-DR, challenging the utility of HLA-DR negativity as a universal release criterion without additional context [46]. Third, trilineage differentiation in vitro remains a useful baseline assay, but it is increasingly recognized as a weak surrogate for in vivo regenerative capacity in most indications, especially those where MSCs are delivered systemically and do not persist long term [47].

2.1.3 Proposed Updates and Consensus Directions

Three converging update streams are most relevant:

The ISCT committee guidance emphasizes (i) retaining "MSC" but annotating tissue source, (ii) using "stromal" unless stringent evidence supports "stem," and (iii) adopting a matrix of functional assays aligned to intended MoA rather than relying solely on minimal identity criteria [7].

Hallmarks and CQAs: a 2025 perspective argues for defining MSC products through hallmarks of identity and potency that can inform surrogate, sensitive CQAs correlated with clinical effectiveness, explicitly motivated by heterogeneous trial outcomes and high-profile late-stage setbacks [48].

A Delphi-driven consensus (2025) proposes a more structured, community-derived definition and reporting guidance for MSC-based therapeutics, responding to broad variability in how "MSC" is defined across trials and publications. Complementary reporting recommendations for autoimmune disease MSC trials emphasize transparent disclosure of manufacturing, characterization, and clinical protocol variables as minimal peer-review criteria [49].

2.2 Tissue Sources of MSCs and Their Comparative Properties

MSCs have been isolated from a wide range of adult and perinatal tissues (Table 2.1). While most sources can yield cells meeting ISCT criteria after culture expansion, source of origin materially influences expansion kinetics, immunophenotype variability beyond the canonical panel, and secretome composition [39].

Table 2.1 Comparing MSC sources across key attributes

MSC source	Typical harvest practicality	Yield (qualitative)	Proliferative capacity (qualitative)	Immunophenotype (core + notable differences)	Secretome/EV highlights (examples)	Common/representative clinical use patterns
Bone marrow (BM–MSC)	Invasive aspirate; long clinical history	Lower starting frequency of stromal progenitors in marrow; expansion required [50]	Often slower growth vs AT/WJ in direct comparisons [39]	Meets ISCT core; HLA-DR can vary with supplements/activation [51]	In some comparisons: strong immunosuppressive potency; secretome may differ in trophic factor abundance vs perinatal sources [39]	Immune indications prominent; off-the-shelf products in some jurisdictions [52]
Adipose (AD/AT–MSC)	Highly accessible; scalable processing	Often high cellularity of stromal compartment; source attractive for manufacturing [53]	Fast growth in many studies; depot-specific differences exist [39]	Meets ISCT core after expansion; passaging shifts phenotypes; CD146-associated functional subsets reported in some contexts [54]	Proteomic secretome distinct from BM and perinatal sources; source-specific payloads [55]	Broad trial landscape across inflammatory/degenerative diseases [56]
Umbilical cord/Wharton's jelly (WJ–MSC)	Non-invasive, perinatal tissue; bankable	High manufacturing attractiveness due to expansion efficiency [57]	Often highest proliferation/shortest doubling time in direct comparisons [39]	Meets ISCT core; differences in non-core markers [39]	High levels of trophic factors in some conditions; distinct secretome composition [39]	Increasing use in pulmonary/inflammatory trials and other indications [58]
Placenta/amniotic sources	Perinatal; non-invasive at birth	Attractive for banking; source-dependent methods vary [59]	Often described as "younger" with high expansion potential [55]	Meets ISCT core after expansion; immunoregulatory interactions emphasized [60]	Secretome proteomics often shows distinct/diverse profiles vs adult sources [55]	Experimental/clinical exploration in inflammatory and regenerative settings [60]

(continued)

Table 2.1 (continued)

MSC source	Typical harvest practicality	Yield (qualitative)	Proliferative capacity (qualitative)	Immunophenotype (core + notable differences)	Secretome/EV highlights (examples)	Common/ representative clinical use patterns
Dental pulp/ oral MSCs (DPSC, etc.)	Minimally invasive with clinical dental access	Moderate; tissue limited but expandable [61]	Variable; depends on donor age and protocols [62]	Meets core MSC phenotype after expansion; distinct transcriptomes/ immunoregulatory properties vs BM in some studies [62]	Source-specific immunomod pathways reported [61]	Tissue repair and oral/maxillofacial regeneration interest; immunomod potential under study [61]
Synovium (S–MSC)	Arthroscopic access; niche relevance for cartilage	Variable; tissue limited but expandable [63]	Often reported as highly chondrogenic; growth varies by donor/disease context [63]	Core phenotype commonly preserved; joint-relevant marker panels increasingly used [63]	Joint-focused trophic/ immunomod effects emphasized; EV approaches expanding [64]	OA/cartilage repair trials and translational studies [65]

2.2.1 *Bone Marrow*

Bone marrow is the historical reference source for MSCs and remains widely used in both research and clinical manufacturing. Comparative studies under clinically relevant conditions show that BM–MSCs can exhibit slower proliferation than adipose- and Wharton's jelly–derived MSCs, while demonstrating strong immunomodulatory activity in T-cell proliferation suppression assays in some settings [39]. Bone marrow MSC manufacturing also illustrates how culture supplements influence phenotype: multi-site GMP analyses show that HLA-DR expression can be present in subsets of clinical-grade BM–MSC batches, prompting re-examination of rigid "HLA-DR negative" release rules under modern xeno-free systems [51].

2.2.2 *Adipose Tissue*

Adipose tissue is attractive due to accessibility and the generally high cellularity of adipose-derived stromal compartments used to derive AD/AT–MSCs, which supports scalable manufacturing and broad clinical exploration [53]. Studies also show that adipose tissue is not uniform: MSC-like stromal cells from different adipose depots can differ in growth rate and angiogenic activity, and CD146 expression may track with higher growth/angiogenic potential in certain adipose-derived MSC populations [54]. A recurring immunophenotyping nuance for adipose-derived MSCs is that marker expression can evolve with passaging, reinforcing the need to define and report passage ranges, culture media, and timepoints for flow cytometry [66].

2.2.3 *Umbilical Cord and Wharton's Jelly*

Umbilical cord tissue, especially Wharton's jelly, is widely used as a perinatal MSC source often associated with high proliferative capacity and rapid expansion. In direct comparisons, Wharton's jelly MSCs (WJ–MSCs) can show shorter doubling times and higher cumulative population doublings than BM–MSCs under matched culture conditions, although time-to-primary culture confluence may differ by source [39]. Secretome studies demonstrate that WJ–MSCs can produce comparatively high levels of trophic factors under certain conditions, and broader proteomic surveys indicate distinct source-dependent secretome profiles [39]. A dedicated review literature also emphasizes the clinical promise and variability of WJ–MSC secretomes and their prospective applications, reinforcing that "WJ–MSC" is not simply a more proliferative substitute for adult tissue MSCs but a biologically distinct starting material [57].

2.2.4 Placenta and Amniotic/Placental Membranes

Placental and amniotic sources provide perinatal MSC populations with immunomodulatory features that have been intensively reviewed and explored experimentally. Reviews emphasize immunoregulatory interactions with T cells, APCs, and broader immune compartments, while primary studies document functional immunomodulation [60]. Proteomic comparisons across MSC sources show placenta-derived MSC secretomes as compositionally distinct and, in some datasets, more diverse than adult tissue MSC secretomes, highlighting how tissue origin can shape "off-the-shelf" trophic/immunomodulatory payloads even when canonical immunophenotypes are met [55].

2.2.5 Dental Pulp and Other Oral Tissue MSCs

Dental pulp stem/stromal cells (DPSCs) and related oral tissue MSCs are increasingly studied for immunomodulation and regeneration. Recent work indicates that oral tissue MSCs possess distinct transcriptomes and immunoregulatory properties relative to adult bone marrow MSCs, implying that oral MSCs should be treated as source-specific products rather than generic MSC substitutes [62].

2.2.6 Synovium- and Joint-Associated MSCs

Synovium-derived MSCs (and closely related infrapatellar fat pad–derived MSCs) have strong relevance for cartilage repair and osteoarthritis. Primary and translational studies increasingly include joint-relevant marker panels and chondrogenic assessments, and recent literature highlights both clinical promise and the need to control for donor disease state, tissue microenvironment, and expansion conditions [63].

2.2.7 Other Sources

Additional MSC sources include amniotic membrane, mucosal tissues, and engineered/iPSC-derived MSC-like populations, each with distinct translational trade-offs (Fig. 2.2). The consistent message across reviews is that tissue source is a primary driver of heterogeneity and must be explicitly controlled and reported [56].

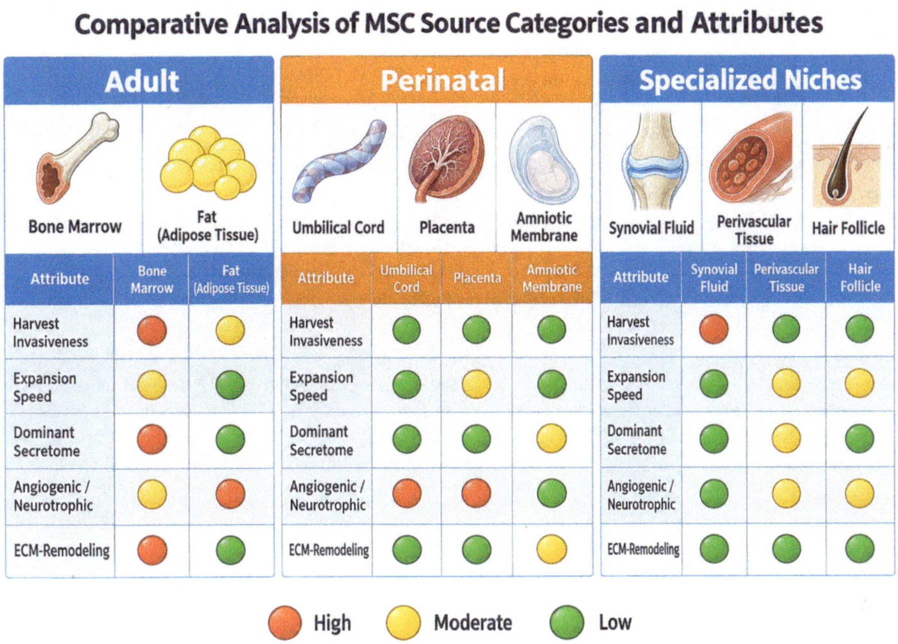

Fig. 2.2 Comparative landscape of mesenchymal stromal/stem cell (MSC) sources: biological origin, harvest burden, expansion kinetics, and secretome profiles. This infographic compares major MSC source categories—adult, perinatal, and specialized niches—across key qualitative attributes relevant to translational applications. Harvest invasiveness is highest for adult sources (particularly bone marrow), while perinatal sources offer minimally invasive or non-invasive collection. Expansion speed is generally faster in perinatal-derived MSCs, with adult MSCs showing comparatively slower proliferation. Specialized niche-derived MSCs exhibit variable expansion depending on tissue context. Dominant secretome themes differ by source: perinatal MSCs are enriched in immunosuppressive and angiogenic factors; adult MSCs frequently exhibit extracellular matrix (ECM) remodeling and structural support profiles; and niche-specific MSCs demonstrate context-dependent secretomes, including neurotrophic and regenerative signaling. Color-coded indicators denote qualitative rankings, providing a practical overview for selecting MSC sources based on therapeutic priorities and manufacturing considerations

2.3 Surface Markers, Characterization, and Quality Control

2.3.1 Canonical Marker Sets and Their Variability

Most studies and manufacturing protocols still begin with the ISCT phenotype: CD73/CD90/CD105 positivity and a lack of hematopoietic/endothelial markers (commonly CD45, CD34, CD14/CD11b, and CD19/CD79a) plus frequently HLA-DR as a negative marker [7] (Table 2.2). However, multiple analyses show that canonical markers are insufficient to capture heterogeneity relevant to potency. Systematic surveys of marker usage in the literature document widespread reliance on CD105/CD90/CD73 but also substantial variability in additional markers, reflecting uncertainty and source-specific practice rather than a single standardized

Table 2.2 Recommended surface marker panels and QC assays

Use case	Recommended flow cytometry marker panel (minimum → expanded)	Key QC assays (identity, safety, potency)	Rationale
Baseline MSC identity (general)	Minimum (ISCT): CD73, CD90, CD105 (positive); CD45, CD34, CD14/CD11b, CD19/CD79a ± HLA-DR (negative) [7]	Morphology + plastic adherence; viability; sterility/mycoplasma/ endotoxin; trilineage differentiation (research identity) [68]	ISCT criteria remain the common baseline but do not predict potency [7]
MSC products expanded in platelet lysate/xeno-free systems	Add context markers: quantify HLA-DR rather than assuming negativity; document HLA-ABC intensity where relevant [46]	Track HLA-DR variability across lots; confirm retention of immunomodulatory function; record media supplement, passage, and activation state [51]	HLA-DR can rise with platelet lysate; binary exclusion may be obsolete without context [46]
Immunomodulatory indication	Expanded phenotype + activation markers as justified; consider CD146/ CD274 (PD-L1) reporting in research contexts [39]	Potency: T-cell proliferation suppression; mixed lymphocyte reaction; secretome/ transcriptome assay matrices; consider IFN-γ licensing in standardized assays [69]	Immunomod potency varies by lot/source; matrix approaches proposed to capture multifunctionality [70]
Secretome/EV-based MSC derivatives	MSC identity as above + EV standards	EV characterization aligned with MISEV2018/ MISEV2023; EV potency assays; dose metrics and controls [5]	EV rigor requires consensus reporting; potency testing is a recognized bottleneck [71]
Advanced harmonized reporting for trials	Report tissue source annotation (BM-MSC, UC-MSC, etc.) and manufacturing parameters; avoid "MSC" ambiguity [7]	Minimal peer-reviewed reporting criteria: product characterization, manufacturing variability, CQAs/potency, dosing/ route, and trial endpoints per guidance [68]	Consensus reporting frameworks aim to reduce trial heterogeneity and improve interpretability [49]

panel [67]. Two high-impact examples illustrate why protocol-dependent variability matters for quality control (QC). Under platelet lysate–based expansion, HLA-DR expression can increase compared with older serum-based assumptions, challenging older release interpretations; multicenter GMP datasets show that HLA-DR+ MSC fractions can retain mesenchymal morphology, identity, in vitro multipotency, and immunomodulatory capacity, requiring more nuanced QC logic than a binary "fail" [46]. CD146 can mark functional differences in MSC subsets in some contexts, reinforcing that subpopulation structure can carry functional meaning [54].

2.3.2 Single-Cell and Spatial/Proteomic Findings: What "Native MSCs" Are—And Why It Matters

Single-cell atlases of bone marrow stroma have transformed how MSC biology is interpreted. Mouse and human single-cell studies define multiple stromal subsets with distinct transcriptional programs and niche functions, emphasizing that culture-expanded "MSC" populations represent a selective, manufacturing-shaped sample of broader stromal ecosystems rather than a single discrete cell type [72]. Human bone marrow mapping combining single-cell transcriptomics with spatial proteomic imaging further refines stromal biogeography and supports a more structured view of niche architecture—important because therapeutic MSCs are often expected to reproduce (or modulate) niche-like functions despite being delivered as dissociated, expanded products [73] (Fig. 2.3).

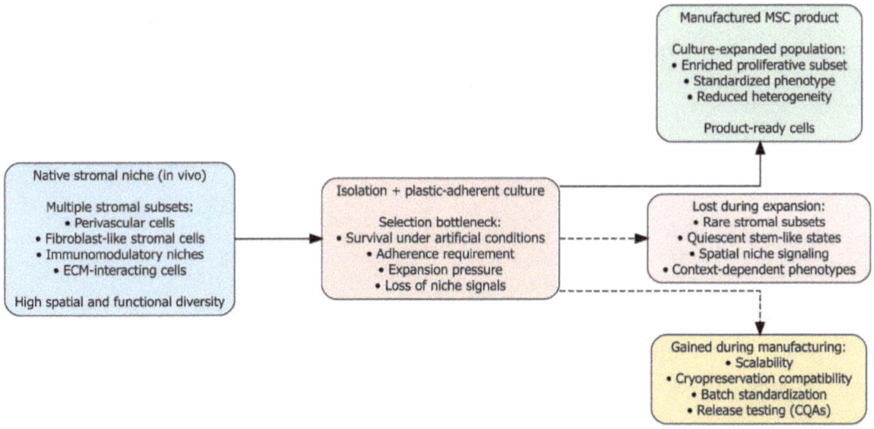

Fig. 2.3 From native stromal niche to manufactured MSC product: selection bottleneck and functional trade-offs. Native stromal tissues contain multiple spatially and functionally distinct stromal subsets shaped by microenvironmental cues. Isolation and plastic-adherent culture impose a strong selection bottleneck, enriching cells that survive and proliferate under ex vivo conditions while eliminating niche-dependent, quiescent, and rare subpopulations. The resulting manufactured MSC product represents a culture-selected subset with improved scalability, standardization, and release test compatibility, but reduced biological diversity compared with the in vivo stromal ecosystem. This transformation underlies key challenges in potency definition, heterogeneity, and reproducibility in MSC-based therapies

2.3.3 Recommended Characterization Approach: Panels Plus Functional QC

A rigorous consistent approach treats MSC characterization as a layered evidence stack (Fig. 2.4):

(i) Identity and purity: flow cytometry with canonical ISCT markers plus additional context markers appropriate to the source/indication [7]
(ii) Manufacturing consistency: viability, sterility, mycoplasma/endotoxin, growth kinetics, and stability/senescence metrics (including passage reporting) [74]
(iii) Potency assays aligned to MoA: direct immunomodulation assays, MoA-linked secretome signatures, and multi-assay "matrices" that better capture MSC multifunctionality [69]
(iv) Advanced profiling (optional but increasingly recommended): transcriptomic/proteomic/EV characterization, especially when secretome/EVs are the intended therapeutic [71]

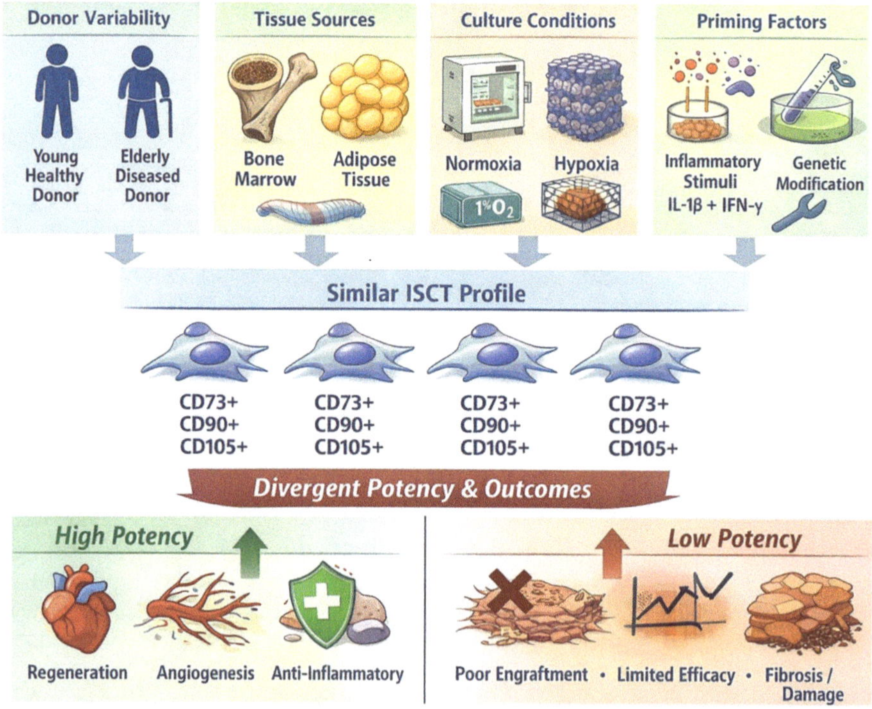

Fig. 2.4 MSC heterogeneity–potency paradox. Schematic illustrating how mesenchymal stromal/stem cells meeting identical ISCT criteria may differ in donor origin, tissue source, culture conditions, and priming status, resulting in divergent potency and clinical outcomes despite similar surface marker profiles

2.4 Differentiation Potential and Lineage Plasticity

Trilineage differentiation (osteogenic, chondrogenic, adipogenic) remains a defining experimental test for MSCs and is incorporated into many source–comparison studies. In comparative experiments under matched culture conditions, MSCs from different tissues can all meet trilineage criteria while differing substantially in growth kinetics, marker expression beyond the canonical panel, and functional immunomodulation [39]. The central translational caveat emphasized in modern reviews is that successful in vitro differentiation does not necessarily predict in vivo tissue formation, engraftment, or long-term functional replacement, particularly after systemic administration where MSC persistence is often short [47]. Thus, differentiation assays should be treated as identity and baseline multipotency evidence, not as a direct measure of therapeutic MoA except in specific local/tissue engineering contexts.

Osteogenesis programs in MSCs are governed by integrated signaling networks that include Wnt, TGF-β/BMP, Hedgehog, and IGF-related pathways and downstream osteogenic transcriptional regulators. Recent pathway-focused reviews emphasize that osteogenic outcomes are highly sensitive to culture conditions, mechanical context, and pathway crosstalk [75].

Chondrogenesis is strongly shaped by TGF-β/SMAD signaling, SOX9 induction programs, and BMP-related modulation, but differentiation outcomes can drift toward hypertrophy depending on protocols. Recent reviews and regenerative engineering studies stress controlling induction conditions and assessing cartilage-relevant matrix markers when MSCs are used for cartilage repair strategies [76].

Adipogenesis depends on transcriptional programs centered on PPARγ and related regulators, and multiple reviews note that adipogenic potential can be source dependent [77]. Claims of differentiation into non-mesodermal lineages persist in the literature, but current mechanistic consensus emphasizes that many observed benefits in such contexts reflect paracrine neurotrophic or immunomodulatory effects rather than stable transdifferentiation. Comparative studies explicitly highlight that secretome differences can produce strong functional readouts even when tissue integration is limited [39].

2.5 Homing, Trophic Effects, and Immunomodulation

2.5.1 Homing and Biodistribution: Why Delivery Route Shapes Mechanism

MSC trafficking after administration is a core determinant of therapeutic MoA (Fig. 2.5). Reviews of MSC homing describe a multistep "adhesion cascade" conceptually analogous to leukocyte extravasation—tethering/rolling, activation, arrest, transmigration, and interstitial migration—mediated by chemokine receptors,

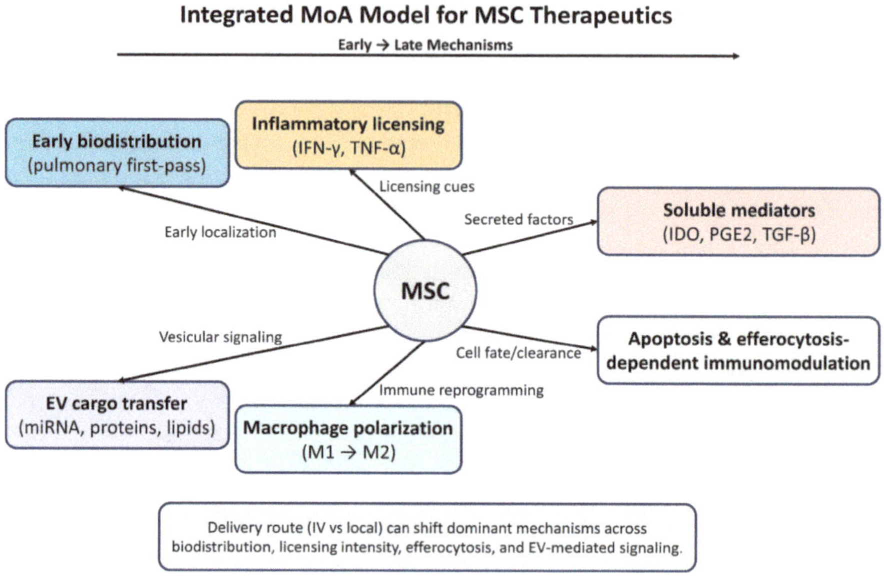

Fig. 2.5 Integrated mechanism-of-action (MoA) model for mesenchymal stromal/stem cell (MSC) therapeutics. Schematic representation of the integrated mechanisms of action underlying MSC-based therapies across early to late phases following administration. After delivery, MSCs undergo early biodistribution, including pulmonary first-pass localization, followed by inflammatory licensing in response to cytokines such as IFN-γ and TNF-α. Licensed MSCs exert therapeutic effects through multiple interconnected pathways, including the secretion of soluble mediators, induction of apoptosis and efferocytosis-dependent immunomodulation, macrophage polarization toward an anti-inflammatory (M2) phenotype, and extracellular vesicle (EV)–mediated cargo transfer (miRNAs, proteins, lipids). These mechanisms collectively contribute to immune regulation and tissue repair. The dominant pathways may vary depending on the route of administration (intravenous versus local delivery), influencing biodistribution, licensing intensity, efferocytosis, and EV-mediated signaling. Abbreviations: MSC mesenchymal stromal/stem cell, MoA mechanism of action, IDO indoleamine 2,3-dioxygenase, PGE2 prostaglandin E2, TGF-β transforming growth factor beta, EV extracellular vesicle

integrins, selectins, and matrix remodeling [78]. However, pharmacokinetic and biodistribution-focused reviews stress that systemically infused MSCs often undergo early lung sequestration ("pulmonary first-pass") and limited long-term retention, with downstream redistribution to organs such as the liver and spleen in many models. This is a major reason the field increasingly interprets MSC efficacy through short-lived immunoregulatory cascades and secretome effects rather than durable engraftment [41].

2.5.2 *Trophic and Paracrine Effects: Secretome as a Major Therapeutic Interface*

A large and growing literature supports MSC action through a complex secretome that includes growth factors, cytokines/chemokines, ECM-modulating proteins, and other bioactive mediators that can promote angiogenesis, limit apoptosis, and reshape local inflammatory states [47]. Source-dependent secretome composition is demonstrated by mass spectrometry profiling comparing adipose, bone marrow, placenta, and Wharton's jelly MSCs, with distinct protein signatures by source even under standardized conditioning approaches [55].

2.5.3 *Immunomodulatory Mechanisms: Soluble Mediators, Licensing, Apoptosis/Efferocytosis, and EVs*

Modern mechanistic reviews converge on several recurring pathways. IDO-dependent immunosuppression is widely discussed as a key immunoregulatory "switch" in MSC-mediated T-cell suppression and broader immune modulation, especially under inflammatory licensing conditions [79]. PGE2/COX-2 axis is repeatedly identified as a central mediator in MSC interactions with innate and adaptive immune cells, including macrophage polarization and suppression of activated lymphocyte responses [80]. TGF-β family mediators contribute to T-cell suppression and Treg support; MSC-derived EVs can carry TGF-β-related immunomodulatory activity as part of their cargo [81]. Inflammatory priming can enhance anti-inflammatory secretome programs (including factors such as TSG-6 in some models), but can also alter immunogenic markers, making priming simultaneously an opportunity and a QC hazard [82]. Apoptosis/efferocytosis-dependent effects: a foundational mechanistic study demonstrates that recipient cytotoxic cells can induce MSC apoptosis and that this process is essential to initiate in vivo immunomodulation in a GVHD model; clinical parallels suggest patient cytotoxic capacity against MSCs may stratify responders vs non-responders in GVHD [40]. MSC-derived EVs/sEVs are increasingly treated as a mechanistically meaningful and potentially more standardized "cell-free" therapeutic. However, field consensus documents emphasize rigorous EV characterization and reporting standards (MISEV2018 and MISEV2023), and ISCT committee work highlights the difficulty of MoA linked potency testing for MSC–sEVs [5].

2.6 Advantages and Limitations of MSC-Based Therapies

2.6.1 Advantages

Updated systematic reviews and meta-analyses in the last decade support that intravascular MSC administration continues to appear generally safe, with adverse event patterns that—while requiring ongoing surveillance for rare/late effects—are broadly reassuring across diverse indications and trial designs [42]. MSCs are widely considered compatible with "off-the-shelf" strategies due to their immunomodulatory properties and manufacturing scalability, supporting large allogeneic banks when CQAs and potency can be controlled [83]. Reviews emphasize that immunomodulatory functions appear to yield more consistent efficacy potential than pure "cell replacement," because durable engraftment is often not required for measurable clinical effects [83].

2.6.2 Limitations and Translational Failure Modes

Multiple analyses identify heterogeneity—by donor, tissue source, manufacturing, culture conditions, passage/senescence, and cryopreservation/thaw status—as a core driver of inconsistent clinical outcomes and reproducibility problems [84]. Regulatory-grade MSC development is repeatedly limited by the difficulty of defining potency assays that are (i) sensitive to meaningful lot-to-lot change, (ii) aligned to intended MoA, and (iii) feasible for routine GMP release. Matrix approaches that combine functional immunoassays with transcriptome/secretome signatures have been proposed as a practical compromise, but harmonization remains incomplete [70]. Reviews and primary studies indicate that cryopreservation can transiently dampen immunomodulatory functions in some settings and that "fresh vs thawed" state should be treated as a controlled variable or mitigated with standardized recovery/reconditioning steps where compatible with clinical workflow [85]. Late-stage trials have underscored that promising early studies do not guarantee phase III success. For example, the 2026 global phase III ADMIRE-CD II trial of darvadstrocel for complex Crohn's perianal fistulas did not meet its primary endpoint at week 24, with no statistically significant difference versus placebo—a result aligned with broader concerns that trial design, endpoints, and placebo response can dominate outcomes [86]. Approvals remain limited and indication specific. A major milestone was the December 18, 2024, approval by the US Food and Drug Administration [41] of remestemcel-L-rknd (Ryoncil) for pediatric steroid-refractory acute GVHD, while other regions have pursued different approval pathways [4].

2.6.3 *Clinical Trial Outcomes: Selected High-Salience Snapshots*

Regarding steroid-refractory acute GVHD, meta-analyses and focused reviews support therapeutic activity in subsets and show substantial trial heterogeneity; regulatory approvals in specific jurisdictions and continued committee-driven standardization initiatives reflect both promise and unresolved variability [87]. Furthermore, about knee osteoarthritis and cartilage repair, evidence remains mixed. Some clinical studies report structural/cartilage signals, but high-quality systematic reviews conclude that intra-articular MSC injection for chronic knee OA pain likely provides little to no improvement in pain/function versus controls, highlighting placebo effects and study quality as key issues [88]. In addition, in COVID-19 ARDS, multiple trials and meta-analyses show heterogeneous results; at least one multicenter randomized trial in COVID-related ARDS did not demonstrate improvements in pulmonary organ dysfunction surrogates, reinforcing indication-specific uncertainty and the importance of standardized potency-linked products [58].

Figure 2.6 illustrates the comprehensive workflow for the development and clinical application of MSC-based therapies, emphasizing the integration of biological, manufacturing, and regulatory considerations across the translational continuum.

The process begins with donor and tissue selection, where strict inclusion and exclusion criteria, informed consent, and infectious disease screening are essential to ensure safety and traceability. This is followed by tissue procurement from various sources such as bone marrow, adipose tissue, umbilical cord (Wharton's jelly), placenta, dental pulp, or synovium, each contributing distinct biological characteristics to the final product. Subsequent primary processing involves washing, enzymatic or mechanical dissociation, filtration, and density-based separation to isolate the cellular fraction of interest. Cells are then subjected to initial culture and plastic adherence, a defining feature of MSCs used for early selection. During expansion under defined conditions, parameters such as media composition, oxygen tension, and passage number are tightly controlled to maintain cell functionality and minimize senescence. In-process controls are implemented throughout manufacturing to monitor growth kinetics, morphology, and contamination. At later stages, identity and purity testing—typically via flow cytometry and viability assays—confirm that the product meets established phenotypic criteria. Crucially, potency testing aligned with mechanism of action (MoA) has emerged as a central requirement, often incorporating immunomodulatory assays, secretome profiling, or multi-assay matrices reflecting the therapeutic intent. Before clinical use, release testing ensures compliance with critical quality attributes (CQAs), including sterility, mycoplasma absence, endotoxin levels, and predefined specifications. The product is then formulated and filled, either as a cryopreserved or fresh preparation, with appropriate dose standardization. Logistics and chain-of-custody systems ensure product integrity during transport, storage, and thawing procedures. Clinical translation culminates in

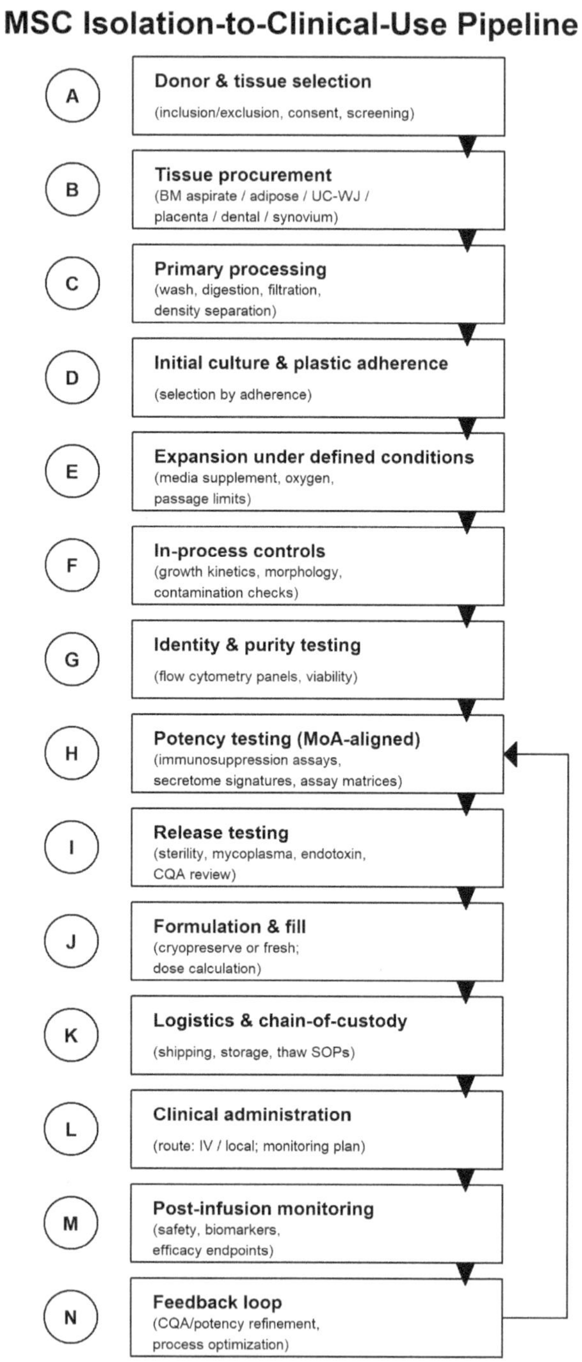

Fig. 2.6 MSC isolation-to-clinical-use pipeline: from donor selection to post-infusion monitoring. Schematic workflow illustrating the end-to-end process for mesenchymal stromal/stem cell (MSC) therapy development and clinical application. The pipeline spans donor and tissue selection, procurement, primary processing, culture expansion, and in-process controls, followed by identity, purity, and mechanism-of-action–aligned potency testing. Subsequent stages include release testing based on critical quality attributes (CQAs), formulation, logistics, and clinical administration, with post-infusion monitoring of safety and efficacy. A feedback loop highlights continuous process optimization and refinement of potency assays and CQAs to ensure product consistency and translational reliability

administration to patients, with route tailored to the indication, followed by post-infusion monitoring of safety, biomarkers, and clinical efficacy endpoints. Importantly, a feedback loop connects clinical outcomes back to manufacturing, enabling continuous refinement of CQAs, potency assays, and process parameters.

Chapter 3
MSC Secretome and EVs

Abstract This chapter focuses on the MSC secretome and its functional components, including extracellular vesicles and soluble mediators. It highlights the complexity of conditioned media and the interplay between EVs and non-vesicular factors in therapeutic outcomes. Methodological considerations, including isolation, characterization, and contamination risks, are discussed in the context of MISEV guidelines. The chapter emphasizes the importance of functional attribution and standardization in interpreting experimental findings.

Keywords Secretome · Extracellular vesicles · Conditioned media · Paracrine signaling · MISEV

A major consolidation across the last decade is the field's movement from ambiguous "exosome" claims toward operationally defined EV preparations with transparent reporting of isolation, characterization, and functional assays [5]. Across MSC sources (bone marrow, adipose, umbilical cord, placenta, gingiva, iPSC derived, etc.), secretome composition is consistently shaped by donor and tissue of origin, microenvironmental structure, and manufacturing variables (2D vs 3D culture, priming, oxygen tension, and serum strategy). Contemporary primary data show measurable differences in cytokine/chemokine abundance and immunomodulatory impact: for example, umbilical cord MSC CM can contain multiple cytokines/chemokines at nanograms per milliliter concentrations and can drive anti-inflammatory macrophage polarization despite a mixed cytokine profile [89].

EVs are now understood as heterogeneous lipid bilayer particles carrying proteins, lipids, and nucleic acids—yet their subclass attribution (exosome vs ectosome/microvesicle) remains difficult to prove in most studies, making nomenclature and marker interpretation a persistent controversy. Recent cargo profiling of clinical-grade EVs indicates that hundreds of miRNAs and proteins can be shared across stem cell products while still retaining strong cell origin specificity, and pathway enrichment repeatedly implicates adhesion/migration, immune regulation, and extracellular matrix remodeling [90].

Therapeutic evidence in the past decade increasingly links "MSC efficacy" to secretome fractions—sometimes dominantly EV mediated, and sometimes

N. M. Mussin, A. Tamadon, *Mesenchymal Stem Cells and Their Extracellular Vesicles*, https://doi.org/10.1007/978-3-032-36437-1_3

requiring synergy between EV and soluble components. These mixed findings are not noise; they reflect biology (multi-component signaling) and methodological variance (dose metrics, isolation method effects on cargo, and contamination controls) [91].

3.1 MSC Secretome Overview

The MSC "secretome" is best treated as a multicomponent product rather than a single entity: it includes (i) soluble proteins (cytokines, chemokines, growth factors, proteases, inhibitors), (ii) lipid mediators and metabolites, and (iii) particulate extracellular material dominated by EVs and other co-isolated extracellular particles. Within the past decade, a recurring theme is that MSC source and manufacturing context frequently explain between-study disagreement as much as disease model biology does [92].

A quantitative illustration of source dependence comes from cytokine/growth factor profiling of CM from multiple MSC tissue sources (adipose, bone marrow, gingiva, placenta, umbilical cord) and culture formats (2D vs 3D). In this work, umbilical cord MSC CM exhibited the highest overall cytokine/growth factor levels and still promoted an anti-inflammatory macrophage phenotype, despite containing prominent immune-recruiting chemokines [89].

At a molecular cargo level, modern profiling of EVs from clinical-grade cell products supports two simultaneous realities: (1) a substantial "shared" repertoire exists and (2) cell-of-origin specificity remains strong. In one head-to-head study, EVs from bone marrow MSCs, umbilical cord MSCs, and heart-derived cells shared 757 miRNAs and 420 proteins, but each source retained distinct highly expressed transcripts and proteins; umbilical cord MSC–EVs showed the greatest uniqueness. Functional/pathway analyses across products converged on adhesion/migration, immune response modulation, platelet-related biology, and RNA/protein processing—implying that many MSC–EV effects are likely emergent network behaviors rather than single-factor mechanisms [90].

Two practical implications follow:

First, unless a tissue source is explicitly specified, "MSC secretome" should be described as a family of related but non-identical products, and the chapter should explicitly treat tissue source and manufacturing parameters as biologically active variables, not minor methods details [93].

Second, "secretome component attribution" (soluble vs EV mediated) should be framed as an experimentally testable question that depends on fractionation strategy, potency assay selection, and dose metric. Direct comparisons of whole CM vs EV fractions vs EV-depleted soluble fractions are therefore central evidence in this field [94].

3.1.1 Conditioned Media, Soluble Factors, and EVs

Conditioned medium is the experimentally captured "supernatant snapshot" of MSC secretion over a defined interval and under defined culture conditions; it therefore contains both soluble mediators and EVs unless EV depleted or fractionated (Fig. 3.1). A recurring finding is that CM often contains high-abundance chemokines/cytokines (sometimes in the ng/mL range), alongside many low-abundance factors that can still be mechanistically important due to receptor sensitivity and pathway amplification. For example, in umbilical cord MSC CM, IL-8, GRO, IL-6, MCP-1, G-CSF, and MCP-3 were each reported at >1 ng/mL under the authors' conditions, and IL-8/GRO was discussed as being in the ~5–20 ng/mL range [89].

The most informative studies for separating "what matters" within CM are those performing within-study fractionation using size- or column-based separation, followed by head-to-head potency assays. One comparative in vitro model (inflamed tenocytes) evaluated whole CM, an EV fraction, and a soluble protein fraction (CM with EVs removed) and reported that (i) CM had the highest protein concentration and particle concentration, (ii) the EV fraction enriched smaller particles, and (iii) whole CM induced the largest gene expression shifts relative to inflamed

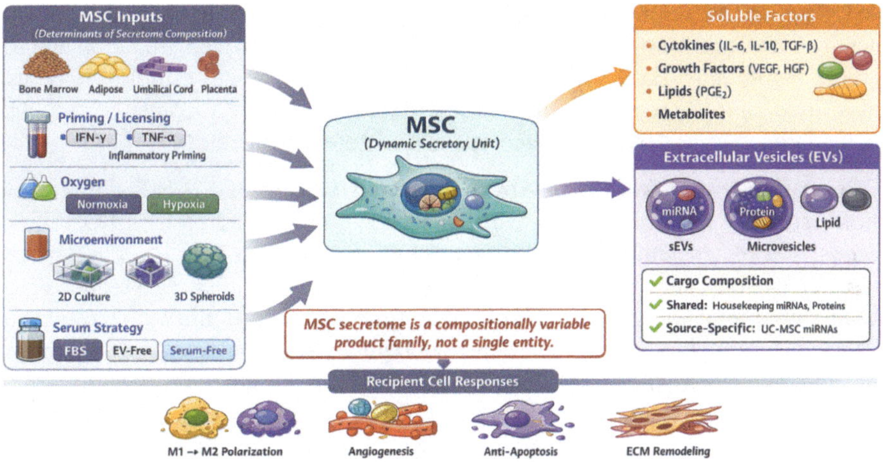

Fig. 3.1 Operational map of the MSC secretome: soluble mediators, EVs, and functional readouts. Mesenchymal stem/stromal cell (MSC) secretome composition is dynamically shaped by upstream determinants including tissue source, priming conditions, oxygen tension, culture microenvironment (2D vs 3D), and serum strategy. These inputs regulate the release of two major output classes: soluble mediators (cytokines, growth factors, lipids, and metabolites) and EVs, which carry proteins, lipids, and nucleic acids such as miRNAs. EV cargo includes both shared components across MSC sources and source-specific signatures observed in clinical-grade profiling studies. Together, these secretome components modulate recipient cell programs including macrophage polarization, endothelial angiogenesis, anti-apoptotic signaling, and extracellular matrix remodeling. The figure emphasizes that the "MSC secretome" represents a compositionally variable product family rather than a single defined entity

controls—supporting the interpretation that soluble and EV components often act synergistically and that "EVs alone" may not reproduce whole secretome activity in some contexts [94].

At the same time, fractionation can also show cases where EVs appear to carry much of the activity. In a renal ischemia–reperfusion model (in vitro OGD/R with tubular HK-2 cells), coculture with MSCs or direct treatment with MSC-derived sEVs inhibited apoptosis, whereas "conditioned medium without sEVs" did not show the same inhibitory effects in the referenced figure description—supporting a more EV-centered mechanism in that setting [91].

3.1.2 Isolation and Characterization Methods for CM and EVs

Because CM and EVs are operational products, isolation/characterization methods are not ancillary; they define what the material is (Table 3.1). The International Society for Extracellular Vesicles' MISEV2018 and MISEV2023 guidance emphasize transparent reporting of (i) source materials and pre-analytics, (ii) separation/concentration methods, (iii) characterization (particle, protein, and markers), and (iv) functional assays with appropriate controls [5].

CM Production (Common Workflow Elements) Most studies (and recommended practice) remove cells/debris by stepwise low-speed centrifugation and filtration before any concentration. Serum strategy is critical: studies may use serum-free intervals or EV-depleted serum to reduce bovine EV contamination, but each approach can alter MSC phenotype and secretory behavior [95].

EV Isolation from CM (Major Approaches and Trade-Offs) Isolation method choice strongly influences yield, purity, and downstream interpretability. A direct comparison of tangential flow filtration (TFF) vs ultracentrifugation (UC)—each followed by size exclusion chromatography (SEC)—showed that TFF reduced processing time for 150 mL conditioned media (minimum ~130 min vs ~210 min for UC) and increased recovery of particles <200 nm by ~11-fold to ~27-fold (method and instrument dependent), while maintaining similar size profiles (majority 100–200 nm; modes ~104–123 nm pre-SEC). The same study cautioned that higher protein recovery does not automatically mean higher purity and used marker/contaminant assessment plus particle-to-protein ratios to evaluate comparability after SEC [96].

EV Characterization (Typical Tool Suite) Field standards emphasize combining particle sizing/counting, morphology (TEM/cryo-EM), and protein markers (multiple positive markers plus negative/contaminant markers) rather than relying on any single assay. MISEV2018 is explicit that no single marker "proves" an EV subtype and that negative controls for co-isolated contaminants are essential [5]. Increasingly, single-EV or multiplex assays are being used to address heterogeneity and marker co-expression at the particle level—an important methodological trend because bulk assays can obscure subpopulation structure [97].

Table 3.1 EV isolation and concentration methods (operational comparison)

Method	Core principle	Typical yield	Typical purity	Scalability/ GMP fit	Cost and time	Key strengths	Key limitations and common co-isolates
Differential UC	Sedimentation by g-force (often with wash)	Medium	Low–medium	Low–medium	High time; specialized equipment	Widely used; no reagents	Co-sedimentation of proteins/ aggregates; operator variability; potential vesicle deformation/ aggregation; purity must be demonstrated [5]
UC + density gradient	Density-based fractionation after UC	Low–medium	High	Low–medium	Very high time/skill	Improved separation from protein complexes/ lipoproteins	Lower recovery; more steps; still not subtype-pure [5]
SEC	Size-based elution through porous resin	Medium	Medium–high	Medium	Medium time; columns	Strong for removing soluble proteins; gentle	Dilution; limited load; often requires reconcentration; resolution depends on column [96]
Ultrafiltration (spin/pressure)	Size cutoff membranes	Medium–high	Low–medium	Medium	Low–medium time	Fast concentration; equipment-light	Membrane adsorption; deformation/shear; co-concentration of protein complexes; requires downstream cleanup for many applications [5]
TFF ± SEC	Continuous-flow filtration; often paired with SEC	High	Medium–high (with SEC)	High	Lower time than UC at scale	Scalable; time-efficient; high recovery	Requires setup expertise; purity still must be validated; downstream SEC often needed [96]
Immunoaffinity capture	Antibody binding to surface proteins	Low	High (for targeted subpopulation)	Low–medium	Medium–high cost	Enrichment of defined antigen-positive EVs	Biased subset; antigen masking; elution artifacts; not ideal for bulk therapeutics [5]

(continued)

Table 3.1 (continued)

Method	Core principle	Typical yield	Typical purity	Scalability/ GMP fit	Cost and time	Key strengths	Key limitations and common co-isolates
Polymer precipitation	Reduced solubility to pellet particles	High	Low	Medium	Low time; reagent cost	High apparent yield; accessible	Strong co-precipitation of proteins/particles; functional results may reflect contaminants unless rigorously controlled [5]
Microfluidic/ lab-on-chip	Custom size/ charge/affinity separation	Variable	Variable	Emerging	Variable	Small samples; rapid workflows	Method-specific bias; limited standardization; translation still developing [71]

3.1.3 Serum Contamination and "Hidden Variables" in CM/EV Work

Within the last decade, the EV field has become unusually explicit that cell culture additives—especially fetal bovine serum (FBS)—can confound both EV analyses and functional assays. An authoritative mini-review documents that FBS-supplemented media can contain EVs and EV-associated cargo that persist even after depletion approaches, affecting in vitro EV studies and recipient cell responses [98]. A related review ("dark side of FBS") summarizes pros/cons and depletion protocols, reinforcing that incomplete removal and batch variability are major reproducibility threats in EV work [99]. Primary work further shows that specific EV depletion protocols can change MSC growth and functional properties, meaning "purity" and "biological equivalence" must be optimized together rather than sequentially [96].

3.2 EV Subclasses and Nomenclature

A central conceptual shift from 2016–2026 is the field's growing insistence that EV nomenclature must reflect evidence, not assumption (Table 3.2). MISEV2018 established that "exosome" should not be used as a default synonym for "small EV," because endosomal origin is usually not proven; instead, the guidelines recommend operational descriptors such as "small EVs" (<200 nm) and "medium/large EVs" (>200 nm), plus explicit reporting of separation methods and markers. The

Table 3.2 EV subclasses and operational nomenclature

EV/particle class (preferred operational term)	Typical size range	Biogenesis route (conceptual)	Commonly reported enrichment markers (not exclusive)	Key caveats for MSC studies
Small EVs (sEVs)	<200 nm (often ~50–150 nm in "exosome-like" preparations)	Often associated with endosomal route, but may include small ectosomes	Tetraspanins and other canonical EV-enriched proteins used as markers; negative markers for contaminants are expected	"Exosome" label generally not warranted without endosomal evidence; sEV prep can still contain non-vesicular co-isolates [5]
Medium/ large EVs (m/ lEVs)	>200 nm (often ~200–1000+ nm depending on method)	Often associated with plasma membrane budding (ectosomes/ microvesicles)	Enrichment markers are less standardized; phenotyping often relies on size plus selected membrane proteins	Size overlap with aggregates/apoptotic bodies in some preparations; requires rigorous imaging and marker controls [71]

(continued)

Table 3.2 (continued)

EV/particle class (preferred operational term)	Typical size range	Biogenesis route (conceptual)	Commonly reported enrichment markers (not exclusive)	Key caveats for MSC studies
Apoptotic bodies/ apoptotic EVs	~0.5–5 μm (broad)	Apoptosis-associated fragmentation/ blebbing	Annexin V binding and apoptotic markers are sometimes used	Easily confused with debris if pre-clearing is inadequate; functional assays require careful controls [5]
EV mimetics/ membrane particles (non-physiologic)	Method defined	Mechanical extrusion/ fragmentation of cells	Method specific	Can yield orders-of-magnitude higher particle numbers per cell than naturally secreted EVs, but differ biologically; should not be conflated with MSC–EVs [5]

MISEV2023 update further reinforces operational naming and emphasizes improved rigor around non-vesicular co-isolates and reporting practices [5].

This matters particularly in MSC biology because therapeutic claims often hinge on subtype explanation ("EVs did X"), while isolation protocols (UC, SEC, precipitation, filtration) enrich overlapping particle populations rather than cleanly separated biogenesis-defined subclasses. Proteomic profiling work explicitly examines how isolation methods can shift the composition of "EV proteome" readouts, complicating mechanistic inference when different labs use different workflows [100] (Fig. 3.2).

3.2.1 Subclass Definitions and Marker Interpretation

Markers Are Enrichment Tools, Not Identity Proofs MISEV2018 recommends demonstrating (at minimum) a set of transmembrane/lipid-bound proteins commonly enriched in EVs, cytosolic proteins expected in EVs, and negative markers for intracellular contaminants. However, it also stresses that no marker set uniquely identifies an "exosome" or "microvesicle" in the absence of biogenesis data [5]. Mechanistic MSC–EV papers often report classical EV markers and sizes consistent with small EVs, but the underlying isolation approach (including precipitation reagents) can affect the interpretation of what was actually isolated [101].

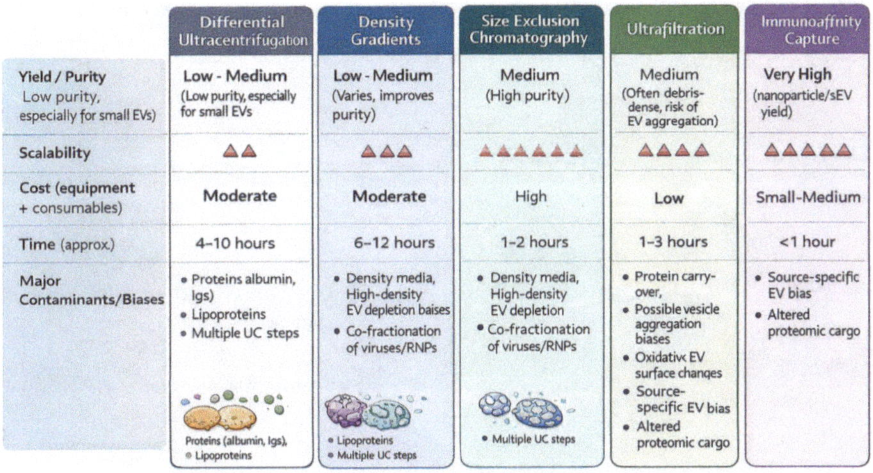

Fig. 3.2 Comparative performance of EV isolation methods for MSC–EV manufacturing and mechanistic studies. Figure comparing differential UC, density gradients, SEC, ultrafiltration, TFF, immunoaffinity capture, and precipitation—summarizing typical yield/purity, scalability, cost, time, and major contaminants/biases. Include quantitative anchors where available and a caution that isolation method can reshape apparent proteomic cargo

3.2.2 Lipids and RNAs as Functional Components

EV membranes are lipid bilayers whose lipid composition influences curvature, stability, and interactions with recipient cells; lipids are also active signaling molecules that can reshape recipient cell responses. Reviews in this period highlight lipid signaling as a key axis in EV biology and emphasize the chemical complexity of EV lipidomes—supporting the principle that "EV function" is not reducible to miRNA alone [102]. On the RNA side, EV cargo profiling in clinical-grade stem cell products demonstrates broad miRNA representation with substantial overlap among products yet strong cell-type specifics among the most abundant transcripts [90]. The principal pathways of extracellular vesicle biogenesis together with the operational nomenclature recommended by MISEV2018 and MISEV2023 are summarized in Fig. 3.3.

3.3 Relationship Between MSC Paracrine Signaling and Therapeutic Action

Across disease categories, MSC secretome-based therapies are typically interpreted through a shared mechanistic "grammar": modulation of inflammation (often via macrophages and lymphocytes), reduction of apoptosis/oxidative injury,

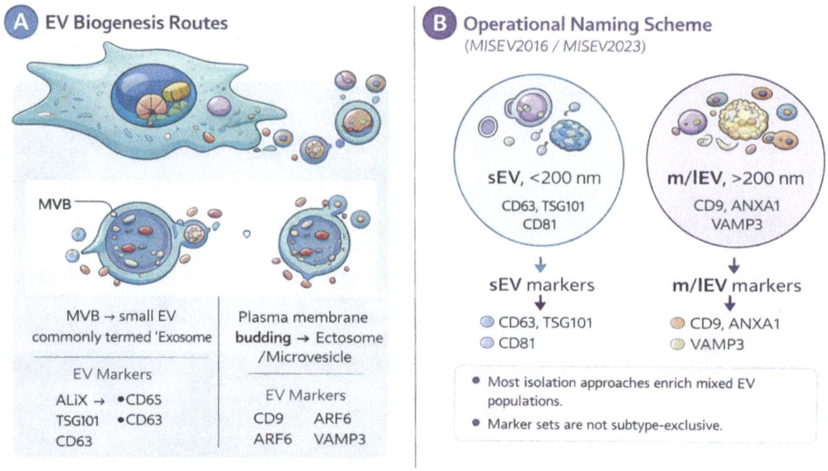

Fig. 3.3 EV biogenesis pathways and the operational nomenclature recommended for MSC–EV studies. A two-panel diagram illustrating (**a**) major EV biogenesis routes, including endosomal multivesicular body (MVB)–derived small EVs, plasma membrane budding–derived ectosomes/microvesicles, and apoptosis-associated vesiculation, and (**b**) the operational naming scheme recommended by MISEV2018 and MISEV2023, including small EVs (sEVs, <200 nm) and medium/large EVs (m/lEVs, >200 nm). The figure emphasizes that most isolation methods enrich mixed EV populations and that marker sets are not subtype-exclusive [5, 109, 116]

Table 3.3 Illustrative comparison of conditioning-dependent changes in MSC-derived EV yield and functional potency

Conditioning	EV yield (particles/10^6 MSC/24 h)	Potency readout at 1 × 10^9 particles/mL
Baseline 2D normoxia	1.0 × 10^10	35
Hypoxia (1% O_2)	2.0 × 10^10	45
Inflammatory priming	1.5 × 10^10	50
3D spheroid culture	1.8 × 10^10	55

promotion of angiogenesis and tissue remodeling, and regulation of fibrotic pathways. What has changed most in the last decade is not this broad framing, but the experimental specificity with which groups now link distinct secretome fractions (or specific EV cargos) to defined molecular targets and functional outcomes [92] (Table 3.3).

3.3.1 Evidence that EVs Can Be Necessary or Sufficient in Some Contexts

Renal Ischemia–Reperfusion Injury A 2024 study reported that human umbilical cord MSC sEVs reduced renal IR injury by delivering miR-100-5p, targeting FKBP5, and modulating AKT phosphorylation signaling; in vivo, animals received 5 × 10^10 particles in 100 μL PBS (tail vein) 6 h before injury. The study also described that sEV treatment inhibited apoptosis in vitro and in vivo and experimentally tied effects to miR-100-5p manipulation [91].

Myocardial Ischemia–Reperfusion/Infarction-Related Injury A 2020 mechanistic study showed that MSC-derived EVs increased cardiomyocyte miR-25-3p, suppressed pro-apoptotic targets (FASL, PTEN), and reduced inflammatory cytokine upregulation; in vivo, 5 μg EVs in 100 μL PBS was injected into the infarct border zone (three sites) 30 min after ligation. This study is also an example of how isolation strategy (precipitation reagent plus exosome-depleted FBS) should be explicitly considered when interpreting mechanistic specificity [101].

3.3.2 Evidence that Whole Secretome Can Outperform EV-Only Fractions

A controlled comparison in inflamed tenocytes found that complete CM produced the strongest gene expression modulation relative to inflamed controls, while both EV and soluble protein fractions had smaller effects (Table 3.3). Quantitatively, CM had ~130 ± 6.2 μg/mL protein and ~ 1.53 × 10^11 ± 2.01 × 10^10 particles/mL, whereas the EV fraction had ~30.4 ± 0.5 μg/mL protein and ~ 1.03 × 10^11 ± 2.04 × 10^10 particles/mL; the EV fraction enriched smaller particles (≤195 nm), while the CM contained more particles >195 nm. This pattern supports a pragmatic conclusion: in some therapeutic contexts, separating EVs from soluble mediators may remove meaningful cooperative biology [94] (Table 3.4).

Local secretome delivery has also shown activity in reproductive tissue repair. In a rabbit model of post-curettage endometrial injury, intrauterine administration of bone marrow MSC–conditioned medium reduced subsequent uterine fibrotic remodeling [103]. Because conditioned medium contains both vesicular and soluble mediators, the study supports local secretome therapy, while also illustrating the need for EV-enriched, EV-depleted, and add-back comparator groups before the therapeutic effect can be attributed specifically to extracellular vesicles.

Table 3.4 Representative therapeutic and mechanistic studies using MSC secretome fractions

Model/indication	MSC source (if specified)	Product evaluated	Dose metric and regimen (as reported)	Main outcome(s)	Proposed mechanistic driver(s)	Key citation
Neonatal hyperoxia-induced bronchopulmonary dysplasia (mouse)	Human umbilical cord MSCs (25 vs 30 weeks GA)	CM and EXO fraction; TSG-6 tests	CM: 10 μg CM protein per mouse IP at PN2 and PN4 EXO: ~2.4 μg EXO protein per mouse, with particle counts reported	Improved lung/cardiac/brain pathology; reduced inflammation, leak, PH, and brain injury markers	Exosome-associated TSG-6 implicated via neutralization/siRNA approaches	[95]
Collagen-induced arthritis and DTH inflammation (mouse)	MSCs (source not specified in excerpt)	EVs vs microparticles	Dose–response; 250 ng EVs or microparticles reported as most efficient dose in DTH; IV dosing in CIA model (timing described)	EVs reduced clinical signs; lymphocyte and B-cell regulatory shifts	EV-associated immunomodulation; EV pellet retained activity	[104]
Renal ischemia–reperfusion injury (mouse + HK-2 OGD/R)	Human umbilical cord MSCs	sEVs (and miR-100-5p modulation)	In vivo: 5 × 10^10 particles in 100 μL PBS via tail vein 6 h before IR injury	Reduced apoptosis, preserved function and histology; mechanistic reversals with miRNA inhibition	miR-100-5p targets FKBP5; AKT phosphorylation axis	[91]
Myocardial ischemia–reperfusion (mouse + in vitro OGD)	Mouse MSCs	"EVs" with miR-25-3p tests	In vivo: 5 μg EVs in 100 μL PBS injected into border zone at 3 sites 30 min after ligation	Reduced infarct size; decreased apoptosis and inflammatory cytokine expression	miR-25-3p targets FASL/PTEN; modulates EZH2/H3K27me3 and derepresses SOCS3/eNOS	[101]
Ischemic stroke (rat MCAO)	Human iPSC–derived MSCs	sEVs	In vivo: 1 × 10^11 particles in 500 μL PBS IV at 4 h after MCAO; mechanistic inhibitor arms included	Reduced infarct volume; enhanced angiogenesis; improved longer-term neurological outcomes	STAT3 activation; reduced autophagy-associated signaling readouts	[105]
Inflamed tenocytes (in vitro)	Equine MSCs	Whole CM vs EV fraction vs soluble protein fraction	Standardized within treatments; EV fractions normalized to 1 × 10^9 EVs/mL; PF normalized to ~72 μg/mL protein; CM normalized by protein (lowest measured)	Whole CM induced the strongest differential gene expression effects compared with inflamed controls	Synergy between EV and soluble mediators; dosing/standardization challenges highlighted	[94]

3.3.3 *Immunomodulation as an Emblematic Paracrine Pathway*

A seminal preclinical arthritis study explicitly compared MSC-derived EVs with microparticles and showed dose-dependent inhibition of T lymphocyte proliferation by both fractions, with EVs being more efficient in suppressing inflammation in vivo (collagen-induced arthritis). The authors also reported that immunosuppressive activity was retained in EV-containing pellets and that 250 ng EVs or microparticles was an effective dose in a delayed-type hypersensitivity model (noted as corresponding to ~2.5 × 10^5 MSC production) [104].

Complementing these EV-focused immunomodulation studies, CM profiling and macrophage assays show that soluble cytokine/chemokine mixtures can push macrophage polarization outcomes and explicitly note that contributions from EVs and lipid mediators remain plausible and are not excluded by cytokine-only profiling. This is a key point: many secretome studies support multi-component immunomodulation even when a single EV cargo is highlighted as "the" mechanism [89].

3.3.4 *Paracrine Signaling in Neurovascular and Neonatal Injury Models*

A 2020 ischemic stroke model in rats reported that human iPSC–derived MSC sEVs promoted angiogenesis and improved longer-term neurological outcomes; dosing was explicitly particle based (1 × 10^11 particles in 500 μL PBS, intravenously 4 h after MCAO), and the authors linked effects to STAT3 activation and reduced autophagy signaling readouts. This is representative of a broader trend: EV therapy papers increasingly include mechanistic pathway tests (pharmacologic inhibitors plus molecular readouts) rather than relying on correlation alone [104].

In neonatal lung disease modeling, a 2018 study in hyperoxia-induced bronchopulmonary dysplasia used serum-free conditioning and evaluated both CM and an "EXO" fraction; dosing was described in multiple formats—10 μg CM protein per mouse or ~ 2.4 μg EXO protein per mouse—and functional experiments implicated the exosome-associated glycoprotein TSG-6 as a key mediator because TSG-6 neutralization/knockdown reduced therapeutic effects. This is an unusually clear example of connecting an EV-associated protein factor to in vivo rescue across multiple organ readouts [95].

3.4 Methodological Standards and Unresolved Controversies

3.4.1 Reporting and Standardization

The EV field has invested heavily (relative to many biomedical subfields) in reproducibility infrastructure (Figs. 3.4 and 3.5). EV-TRACK was created to centralize methodological metadata and promote transparent reporting; subsequent work added tools to integrate checklists and summaries into publications. For MSC secretome/EV chapters, this matters because many contradictions in "MSC–EV efficacy" can be explained by incomplete reporting of key variables (media formulation, depletion, isolation method, characterization depth, and dose metric) [106].

MISEV2018 and MISEV2023 provide the most influential consensus scaffolding for what "adequately characterized EVs" means in practice: multiple orthogonal characterization measures, explicit controls for contaminants, and careful operational naming of EV preparations. These guidelines should be treated as central chapter-defining references [5].

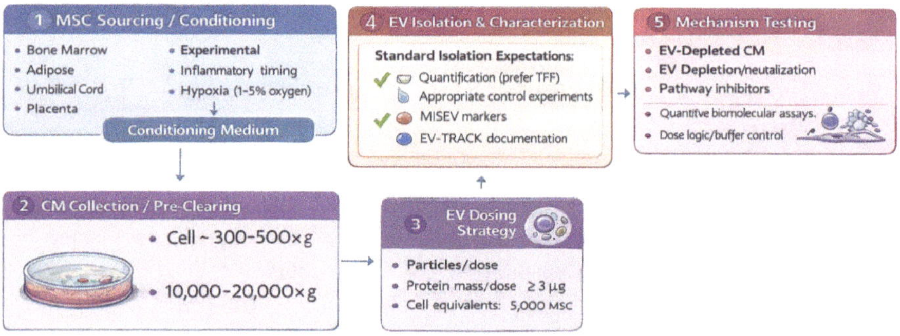

Fig. 3.4 From MSC culture to a defensible therapeutic claim: secretome/EV production, QC, and mechanism testing. Flowchart emphasizing (i) MSC sourcing and culture variables, (ii) CM collection and pre-clearing, (iii) EV isolation and characterization aligned to MISEV/EV-TRACK expectations, (iv) dosing strategy (particles vs protein vs cell equivalent), and (v) mechanism testing using depletion/neutralization, EV-depleted CM, and pathway inhibitors

Fig. 3.5 (continued) robust and defensible therapeutic claims. The process begins with MSC sourcing and culture conditioning variables (tissue origin, donor characteristics, priming, oxygen tension, and 2D/3D systems), followed by conditioned medium (CM) collection and pre-clearing. EV isolation and characterization are then performed using methods aligned with the International Society for Extracellular Vesicles guidelines (MISEV2018/MISEV2023) and reporting standards such as EV-TRACK. Subsequent steps include definition of dosing strategies (particle number, protein mass, or cell equivalent) and functional validation assays. Mechanistic attribution is established through depletion/neutralization approaches, comparison with EV-depleted CM, and pathway perturbation experiments. The workflow emphasizes that rigorous methodological transparency and multi-level validation are essential for linking MSC-derived products to reproducible biological effects and clinical translation

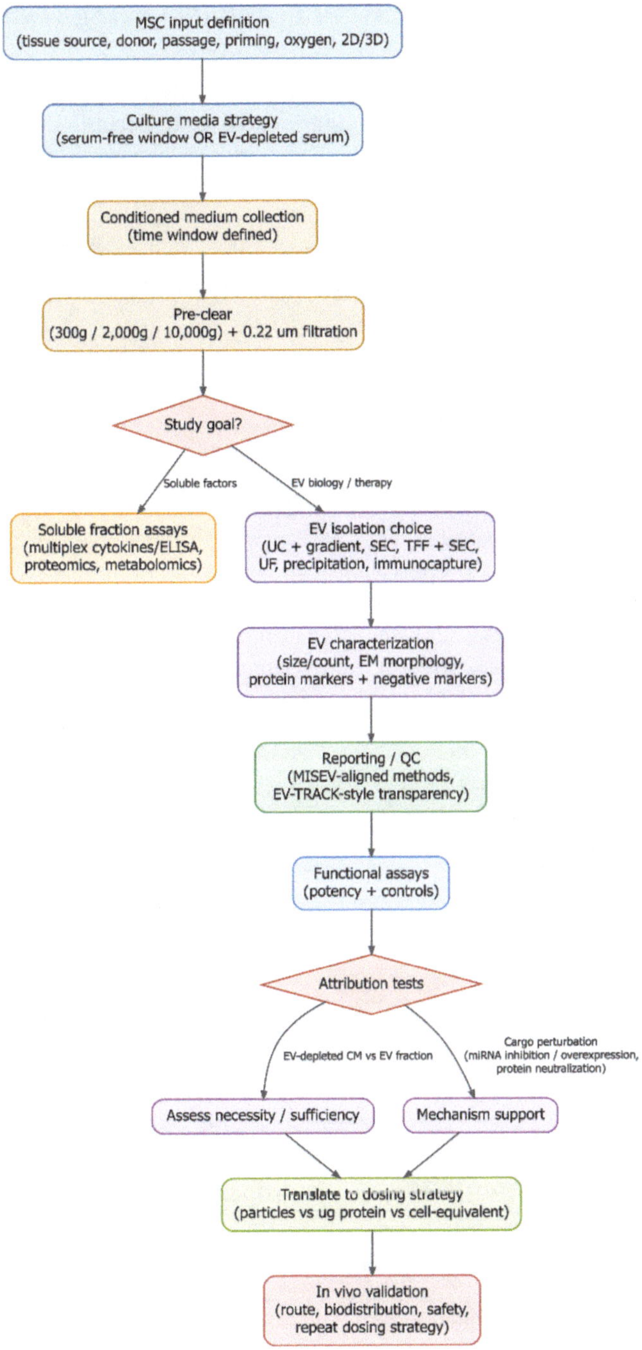

Fig. 3.5 From MSC culture to a defensible therapeutic claim: secretome/EV production, quality control, and mechanism testing workflow. Workflow illustrating the sequential steps required to translate mesenchymal stromal/stem cell (MSC) secretome and extracellular vesicles (EVs) into

3.4.2 *Contamination and "What Is Actually Being Delivered"*

A persistent unresolved issue in MSC secretome studies is disentangling EVs from co-isolated proteins, protein aggregates, and serum-derived particles. Reviews specifically highlight that FBS-derived EVs can confound cultured-cell EV work and that depletion is imperfect and protocol dependent [98]. Meanwhile, method-comparison proteomics emphasizes that isolation workflow affects observed EV protein cargo, raising the possibility that some "mechanistic cargos" are, in part, workflow artifacts unless validated across orthogonal methods [100].

3.4.3 *Dose Metrics and Potency Assays*

There remains no universally accepted standard for dosing MSC–EV therapeutics. Recent primary work illustrates why: investigators report doses in protein mass, particle number, and/or "cell-equivalent" production, sometimes in the same study [95]. This heterogeneity makes cross-study comparisons difficult and can change conclusions in "EV vs CM" debates, since protein-based and particle-based normalization can privilege different components of a mixed preparation [94] (Fig. 3.6).

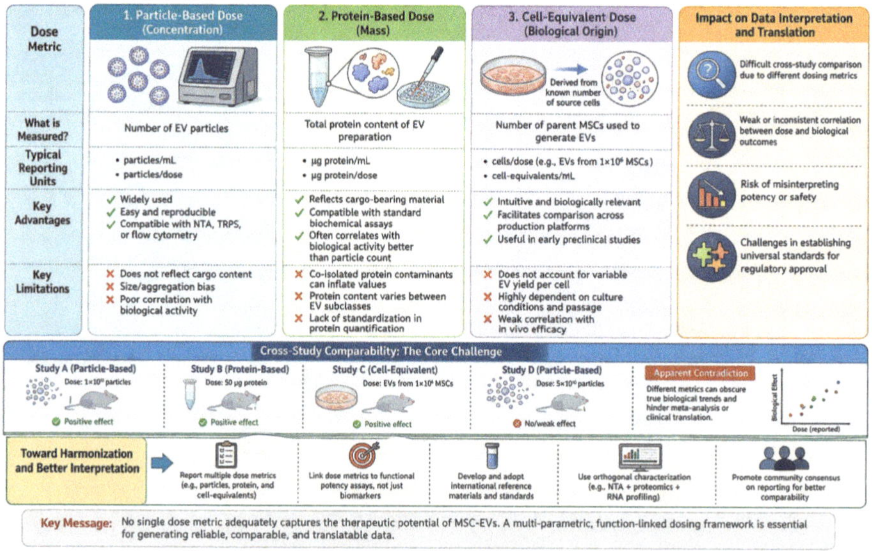

Fig. 3.6 Divergence of EV dose metrics across studies. Comparison of particle-based, protein-based, and cell-equivalent dosing strategies and their impact on interpretability, potency correlation, and cross-study comparability

3.4.4 *Scalability for Translation*

For translation, scalable isolation approaches and robust QC pipelines are pivotal. A direct methods comparison demonstrates that TFF combined with SEC can greatly improve sEV recovery and reduce processing time compared with UC, suggesting an increasingly plausible route to larger-scale production—while emphasizing that purity must still be demonstrated with appropriate controls [96]. Reviews in this era frame manufacturing as a central bottleneck for EV therapeutics and argue that "process defines product" for EV-based medicinal development [107].

3.5 Critical Synthesis and Translational Interpretation

The current evidence suggests that MSC secretome and MSC–EV preparations should not be interpreted as interchangeable therapeutic products. Although EVs can reproduce several MSC-associated paracrine effects, the relative contribution of EVs, soluble proteins, lipids, metabolites, and co-isolated extracellular components remains context dependent. Therefore, the central translational question is not whether MSC–EVs are "active," but under which biological, manufacturing, and disease-specific conditions EV-enriched preparations provide advantages over whole conditioned medium or vesicle-depleted fractions.

A critical interpretation of the field indicates that three factors repeatedly determine the reliability of therapeutic claims: first, the degree to which the active fraction has been experimentally attributed using depletion, add-back, or fractionation approaches; second, whether the dose has been reported in biologically interpretable units; and third, whether the isolation method preserves functional cargo while reducing confounding contaminants. Future studies should therefore move beyond descriptive comparisons and adopt mechanism-linked, side-by-side experimental designs that compare whole secretome, EV-enriched fractions, and EV-depleted fractions under standardized potency assays.

Chapter 4
MSC-Derived EV Biogenesis, Cargo, and Mechanisms of Action

Abstract This chapter explores the biogenesis, molecular cargo, and mechanisms of action of MSC-derived extracellular vesicles. It outlines key pathways involved in EV formation, including ESCRT-dependent and ESCRT-independent mechanisms, and examines how cellular state influences EV composition. Functional cargo, including proteins, lipids, and RNAs, is linked to therapeutic effects such as immunomodulation and tissue repair. Challenges related to cargo heterogeneity, uptake mechanisms, and in vivo biodistribution are also discussed.

Keywords EV biogenesis · Cargo profiling · miRNA · Mechanisms of action · Biodistribution

MSC–EVs are heterogeneous lipid bilayer particles that can recapitulate many MSC "paracrine" effects—repair signaling, pro-angiogenic activity, anti-apoptosis, and immunomodulation—without the replicative and engraftment-related risks inherent to live cell administration. The field has converged on a view that "MSC–EVs" is best treated as an operational product class (defined by parent cell identity and standardized isolation/characterization) rather than a single mechanistically uniform entity, because EV preparations typically include multiple vesicle subtypes and co-isolated non-vesicular components [5].

The biogenesis of MSC–EVs spans (i) endosomal multivesicular body (MVB)–derived small EVs often termed "EVs," (ii) plasma membrane–budded microvesicles (also called ectosomes), and (iii) apoptotic bodies generated during programmed cell death. Core biogenesis machinery includes ESCRT components, tetraspanins, lipid remodeling pathways, and trafficking/fusion regulators such as Rab GTPases; MSC culture state (2D vs 3D, density, hypoxia/inflammation "licensing," senescence) measurably modifies EV yield and cargo composition, complicating cross-study comparability but also creating an engineering lever for potency [108–110].

MSC–EV cargo comprises proteins, lipids, RNAs (miRNAs, other small RNAs, mRNAs), and DNA fragments. Quantitative "omics" indicates enrichment of canonical EV markers alongside parent cell– and condition-specific molecules, including immunoregulatory checkpoint ligands and regulatory miRNAs that can be causally linked to functional phenotypes via gain/loss experiments. However,

N. M. Mussin, A. Tamadon, *Mesenchymal Stem Cells and Their Extracellular Vesicles*, https://doi.org/10.1007/978-3-032-36437-1_4

cargo profiles are sensitive to isolation method, which can change both detected proteins and inferred pathways [111].

In vivo biodistribution after systemic delivery is dominated by reticuloendothelial sequestration (liver/spleen) with smaller fractions reaching injured targets; quantitative PET of radiolabeled MSC–EVs in a spinal cord injury model illustrates high liver and spleen uptake with detectable, injury-enhanced spinal cord signal. Disease context can redirect EV homing and/or uptake [112, 113].

Mechanistic evidence is strongest where the field demonstrates (a) defined EV characterization and dose metrics, (b) cargo necessity/sufficiency, and (c) replicate functional assays. Causal links exist for PTEN–PI3K/AKT survival signaling via exosomal miR-21 in intervertebral disc degeneration, PD-L1-dependent suppression of T-cell receptor signaling via PD-L1 + MSC–sEVs, and miR-125a-3p-mediated attenuation of neutrophil extracellular trap (NET) formation in spinal cord injury. Nonetheless, major gaps remain in (i) distinguishing exosome vs microvesicle contributions in typical "MSC–EV" preparations, (ii) standardizing potency unit rigorously, and (iii) reconciling in vitro uptake and dose–response with in vivo pharmacokinetics and tissue-level exposure [114].

4.1 Approach and Conventions

This chapter is written to support a rigorous "Chap. 4" for the book titled *MSC-Derived EV Biogenesis, Cargo, and Mechanisms of Action*. The evidence base was prioritized toward 2016–2026 primary research and authoritative reviews indexed in PubMed/PMC, emphasizing mechanistic studies with defined EV characterization and causal perturbation experiments. A small number of foundational pre-2016 papers are referenced only where they remain central to mechanistic understanding and are explicitly treated as background rather than the primary evidentiary basis [115].

Terminology and Reporting Standard Consistent with community guidance, "extracellular vesicles (EVs)" is used as the default term unless the biogenetic route is directly demonstrated; the term "exosome" is reserved for cases with strong evidence of endosomal origin and "microvesicle/ectosome" for plasma membrane budding. This reflects the consensus that size-based fractions alone do not prove biogenesis [5, 109].

Evidence–Strength Rubric Used Throughout Mechanistic strength is described as high = cargo necessity/sufficiency shown plus functional assays; moderate = strong association plus partial perturbation; and low = descriptive profiling or correlative functional assays without cargo linkage. This rubric is aligned with how major EV standards emphasize controls, purity/identity assays, and functional specificity [5].

Citation Style Note You requested inline author–year citations; no other citation style was specified. This chapter therefore uses narrative author–year in-text references and does not format a full bibliography here, since the target book's editorial style is unspecified [5].

4.2 Biogenesis and Release of MSC-Derived EVs

4.2.1 Conceptual Map of MSC–EV Biogenesis Pathways

MSC–EV production can be schematized as three overlapping routes (Fig. 4.1): MVB-derived small EVs, plasma membrane–shed microvesicles/ectosomes, and apoptotic bodies. In practice, many isolation workflows (differential centrifugation, filtration-based enrichment) yield mixtures of these vesicle types plus non-vesicular particles; thus, the "MSC–EV product" is often best treated as a composite [5, 109, 116].

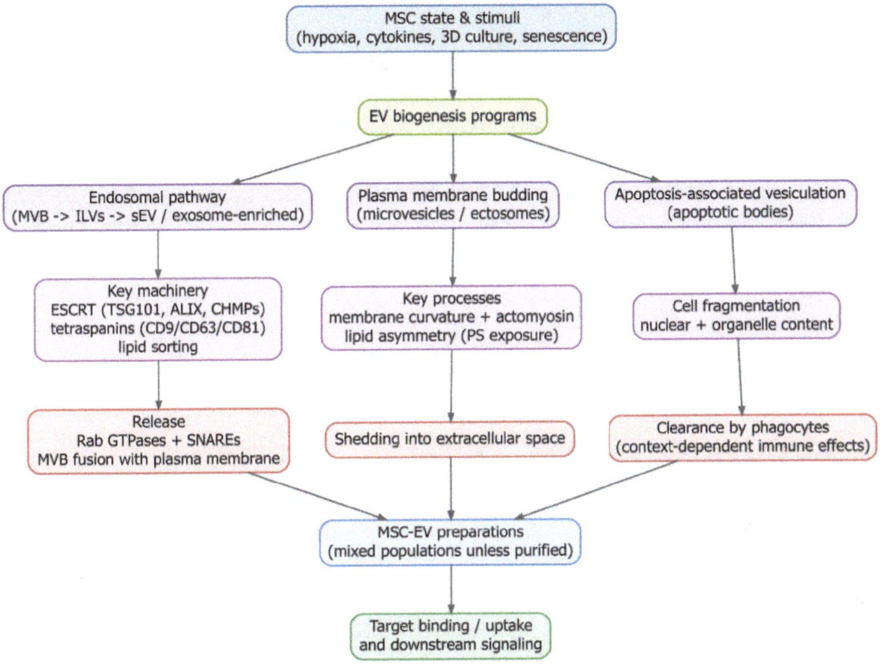

Fig. 4.1 MSC-derived extracellular vesicle biogenesis pathways and points of regulation. Schematic representation of mesenchymal stromal/stem cell (MSC)–derived extracellular vesicle (EV) biogenesis pathways and regulatory mechanisms. MSC state and environmental stimuli—including hypoxia, inflammatory cytokines, three-dimensional (3D) culture conditions, and cellular senescence—modulate EV biogenesis programs. Three principal pathways are illustrated: (i) the endosomal pathway, involving multivesicular body (MVB) formation and release of intraluminal vesicles (ILVs) as small EVs (commonly termed EVs), regulated by ESCRT machinery, tetraspanins, and lipid sorting mechanisms; (ii) plasma membrane budding, generating microvesicles/ectosomes through cytoskeletal remodeling and membrane lipid asymmetry; and (iii) apoptosis-associated vesiculation, producing apoptotic bodies containing nuclear and organelle components. These vesicles are typically co-isolated in standard preparations, highlighting that size-based isolation alone does not establish vesicle origin. The resulting heterogeneous MSC–EV populations interact with recipient cells through binding, uptake, and downstream signaling pathways. Adapted conceptually from the International Society for Extracellular Vesicles guidelines and foundational EV biology studies

4.2.2 Key Molecular Machinery Relevant to MSC–EV Generation

ESCRT-Dependent and ESCRT-Associated Mechanisms Endosomal budding into MVBs (forming intraluminal vesicles) is commonly linked to ESCRT complexes, with ESCRT-associated proteins frequently detected in EV proteomes and used as part of marker panels—although their presence is not exclusive or sufficient proof of "EVs" [5, 109, 111].

ESCRT-Independent Mechanisms and Tetraspanin Platforms Tetraspanins (CD9, CD63, CD81) are highly prevalent EV membrane proteins and are central both to marker-based identity assays and to mechanistic models of membrane microdomain organization, cargo clustering, and potential targeting [5, 109].

Rab GTPases and Intracellular Trafficking Vesicle trafficking and fusion dynamics (MVB–plasma membrane fusion, endosomal recycling) are broadly regulated by small GTPases (Rab family) and associated trafficking machinery. In MSC contexts, Rab-associated regulation is often invoked when culture perturbations or cell-state changes modify EV secretion rates; however, direct causal mapping of specific Rab nodes to MSC–EV output remains less standardized than in cancer exosome models [109].

Microvesicle/Ectosome Budding and Membrane Remodeling Microvesicle formation from the plasma membrane involves localized membrane curvature, cytoskeletal remodeling, and altered lipid asymmetry. Conceptually, this route is distinct from MVB release; practically, conventional MSC–EV preparations commonly co-isolate ectosomes and sEVs unless explicitly fractionated [5, 116].

4.2.3 Regulation by MSC State and Stimuli

3D Culture and Non-adherent Morphology as Yield Modulators (High Evidence for Yield Effects) A well-controlled 3D spheroid study demonstrated that human bone marrow MSCs in 3D spheroids produce substantially more EVs than 2D monolayer cultures and quantified exosome yield (μg per 1 × 10^5 cells) across spheroid sizes and culture conditions. The authors' mechanistic interpretation favored the non-adherent, rounded morphology as a driver, with spheroid size/density modifying efficiency [108].

Microenvironment-Induced Cargo Shifts (Moderate Evidence; Strong on Composition Differences, Variable on Functional Gains) Proteomic composition of MSC–EVs differs between 2D and 3D microenvironments, and these differences can map to divergent in vitro activity profiles. A 2D vs 3D comparison reported distinct EV protein composition depending on culture microenvironment and cautioned that 3D-derived EVs did not universally enhance "typical" MSC–EV properties across assays, emphasizing the need for potency testing rather than assuming yield correlates with efficacy [110].

Inflammatory Licensing and Immunoregulatory Phenotype (Moderate-to-High Evidence in Specific Mechanistic Models) Cytokine priming is a commonly used strategy to enhance immunomodulatory activity of MSC secretomes, including EV fractions. In a priming study of stromal cell secretomes, combined IFN-γ/TNF-α priming was proposed as a strategy to enhance immunomodulatory capacity [117]. At the level of defined EV cargo, PD-L1 expression on MSC–sEVs is functionally important in a T-cell suppression model; the same study discusses inflammatory cytokines as a regulator of PD-L1 release and licensing [118].

4.2.4 Critical Appraisal, Gaps, and Future Directions for Biogenesis

A central methodological limitation is that MSC–EV "biogenesis claims" frequently extrapolate from general cell biology models rather than directly demonstrating subcellular origin in the specific MSC system under study. Current standards encourage reporting multiple complementary markers and controls to support EV identity/purity and to avoid over-claiming "exosome" specificity [5, 111]. Key gaps include (i) systematically separating MSC exosome-like sEVs from ectosomes with matched functional readouts, (ii) mapping how specific priming conditions rewire cargo sorting (not merely yield), and (iii) bridging manufacturing-oriented yield changes to mechanistic potency correlates suitable for release assays [5, 108, 110].

4.3 Molecular Cargo of MSC-Derived EVs and Methods for Profiling

4.3.1 Cargo Classes and Functional Logic

MSC–EV cargo can be organized into proteins (surface and luminal), lipids (structural and signaling), RNAs (miRNAs, other small RNAs, mRNAs), and DNA fragments (genomic and mitochondrial). Each cargo class has distinct evidence for (a) active sorting vs passive encapsulation, (b) stability in circulation, and (c) likelihood of being functionally delivered at sufficient stoichiometry [5, 109, 119] (Table 4.1).

Table 4.1 Cargo types, functions, and common profiling approaches

Cargo type	Representative MSC–EV examples	Dominant functional hypotheses	Core detection/profiling methods	Key artifacts/controls
Proteins (membrane/luminal)	Tetraspanins (CD9/CD63/CD81), ESCRT-associated proteins (TSG101/ALIX), immunoregulatory PD-L1 in specific products	Receptor–ligand signaling at cell surface; enzymatic activity; modulation of immune checkpoints; intracellular pathway tuning after uptake	Western blot, bead-assisted flow cytometry, targeted ELISA, LC–MS/MS proteomics	Co-isolated protein aggregates; method-dependent proteome shifts; need negative markers per minimal standards
Lipids	Cholesterol, sphingomyelin/ceramide-related lipids; lipid microdomain organization; enrichment of ceramide/sphingomyelin reported in ASC–sEVs	Membrane stability; uptake modulation via raft domains; bioactive lipid signaling	Lipidomics (LC–MS/MS, shotgun lipidomics); targeted assays	Lipoprotein contamination; dye-labeling artifacts in biodistribution studies
miRNAs and small RNAs	miR-21 (survival signaling), miR-125a-3p (NET regulation), broad small RNA landscape in MSC–EVs	Post-transcriptional gene regulation in recipient cells; pathway reprogramming	Small RNA-seq, RT–qPCR, ddPCR	Low copy number/stoichiometry concerns; need spike-ins and library bias controls
mRNAs	Transcript fragments associated with repair phenotypes; mRNA presence reported across EVs	Translation in recipient cells or innate sensing	RNA-seq, RT–qPCR	RNase contamination, extracellular RNA complexes
DNA fragments	mtDNA enrichment in ectosome fraction; EV-associated DNA in general	Innate immune activation (DAMP-like); biomarker potential	qPCR for mtDNA/nDNA; sequencing-based assays	Distinguish intraluminal DNA from surface-bound DNA; nuclease protection assays

4.3.2 Quantitative Signals and Notable Cargo Examples

Proteins: Marker Panels Versus Functional Proteins Many MSC–EV studies rely on canonical marker panels (tetraspanins, HSP70, TSG101) for identity, as demonstrated in mechanistic biodistribution work that detected CD9/CD63/CD81 and internal proteins including HSP70/TSG101, with negative controls used to assess contamination in some systems [113]. Functionally specific proteins can be therapeutically decisive: Wharton's jelly MSC–derived sEVs enriched in PD-L1 suppressed T-cell receptor signaling (pZAP70–NFAT) and required PD-L1 for activity, with antibody neutralization and PD-L1 knockout experiments supporting causality [118].

Lipids: Beyond "Membrane Material" Reviews emphasize that EV lipid composition differs from parent membranes and can influence both biogenesis and uptake; sphingolipids and cholesterol-rich domains are frequently discussed as functional determinants [120]. A focused stem cell EV study reported enrichment of free fatty acids, ceramide, and sphingomyelin in adipose-derived MSC small EVs compared with donor cells, supporting the view that lipid cargo is actively shaped and may contribute to barrier/epidermal signaling phenotypes [121].

miRNAs and Small RNAs: Functionally Tractable Cargo with Causal Toolkits Small RNA sequencing has been used to map comprehensive small non-coding RNA landscapes in MSC-derived EVs, providing a discovery basis for selecting candidate functional miRNAs for perturbation [119]. At the mechanistic level, miR-21 carried by MSC–EVs was shown to transfer into nucleus pulposus cells and suppress apoptosis through PTEN targeting and PI3K/AKT activation; in vivo, intradiscal injection alleviated disc degeneration, and downregulation of exosomal miR-21 dampened benefit—supporting cargo causality rather than simple association [114]. In spinal cord injury, an MSC exosome study combined functional outcomes with a specific miRNA mechanism: the miR-125a-3p mimic diminished NET formation, and the inhibitor reversed exosome effect, supporting miRNA-mediated action [112].

DNA and Mitochondrial Components: Emerging Mechanistic Ambiguity A 2022 analysis comparing ectosome-like vs exosome-like fractions from MSCs quantified mitochondrial DNA (mtDNA) using qPCR (multiple genomic regions) and reported ~30-fold mtDNA enrichment in ectosomes versus cells, with relative depletion in EVs. The authors emphasized that mtDNA presence does not prove intact functional mitochondria in EVs and framed mitochondrial components as potential DAMP-like signals [122].

4.3.3 Methods for Cargo Profiling and Their Interpretive Risks

Proteomics (LC–MS/MS) and the Isolation-Method Problem Proteomic profiling of MSC-derived EVs is particularly sensitive to the isolation workflow. A 2024 proteomic analysis explicitly evaluated how isolation methods alter detected protein cargo, underscoring that "cargo differences" across studies can reflect processing, not biology, unless methods are harmonized and controlled [100]. A proteomic comparison work frequently referenced in EV biology also demonstrated that EV subtypes can have distinct proteomes, reinforcing that "EV preparation" is not inherently monolithic [111].

RNA Profiling (Small RNA-Seq, mRNA-Seq, Targeted RT–qPCR) Deep sequencing can define miRNA abundance patterns and support pathway enrichment analyses; a cardiac repair paper used miRNA sequencing to compare MSC–EVs and parental MSCs and reported read-based expression metrics (RPKM) and then mapped differentially expressed miRNAs to PI3K–AKT and mTOR pathway enrichment [123]. Across RNA cargo studies, controls are essential: spike-in standards, library-bias correction, and nuclease protection assays to distinguish encapsulated from surface-bound or co-isolated extracellular RNA. These controls are aligned with minimal EV characterization standards for functional claims [5, 119].

Lipidomics Lipidomics requires particularly stringent control for lipoprotein contamination and for downstream dye-labeling artifacts, because lipid dyes can label non-EV lipid structures. Reviews emphasize that lipid species can influence EV lifecycle, but translational interpretation requires careful sample purity and orthogonal validation [113, 120]. Table 4.2 emphasizes molecules with (i) frequent reporting in MSC–EV literature and/or (ii) strong mechanistic linkage to function in primary studies (Fig. 4.2).

Table 4.2 Selected MSC–EV cargo molecules with functions and detection methods

Cargo molecule/class	EV association type	Canonical or proposed function in MSC–EV context	Typical detection method(s)	Evidence strength
PD-L1	Surface protein (sEV)	T-cell suppression via PD-1 engagement; modulation of TCR signaling (pZAP70–NFAT)	Immunoblot; functional neutralization; genetic knockout	High [118]
CD9/CD63/CD81	Surface marker proteins	Marker/identity; may contribute to targeting microdomains	Immunoblot; bead-flow; EM-associated labeling	Moderate (markers, not necessarily functional) [113]
HSP70, TSG101	Luminal/associated proteins	Marker/biogenesis associated; stress response interpretation	Immunoblot; proteomics	Moderate [113]

(continued)

Table 4.2 (continued)

Cargo molecule/class	EV association type	Canonical or proposed function in MSC–EV context	Typical detection method(s)	Evidence strength
miR-21	miRNA	Anti-apoptosis/survival via PTEN targeting → PI3K/AKT activation; tissue repair	RT–qPCR; functional mimic/inhibitor; in vivo injection	High [114]
miR-125a-3p	miRNA	Reduced NET formation; anti-inflammatory effect in SCI model	miRNA selection + mimic/inhibitor; functional assays; PET biodistribution	High [112]
miR-27a-3p	miRNA	Alveolar macrophage M2 polarization; reduced inflammation in ALI	RT–qPCR; macrophage assays; in vivo lung injury outcomes	Moderate–high [124]
miR-24, miR-29	miRNA	Anti-apoptosis/antifibrosis pathway associations in MI model	miRNA-seq; pathway enrichment; in vitro dose–response assays	Moderate [123]
mtDNA fragments	DNA	Potential DAMP signaling; biomarker; unclear energetic contribution	qPCR across mtDNA regions + mtDNA/nDNA ratio	Moderate (quantitative presence; functional ambiguity) [122]
Ceramide/sphingomyelin/FFAs	Lipids	Potential effects on membrane rigidity, uptake propensity, and signaling	LC–MS/MS lipidomics	Moderate (composition changes; function often indirect) [121]

4.4 Target Cell Uptake, Targeting Determinants, and Biodistribution

4.4.1 Uptake Mechanisms at the Target Cell Interface

EV uptake is mechanistically plural: clathrin-independent endocytosis, macropinocytosis, phagocytosis, and other cholesterol-dependent pathways can contribute, with pathway predominance depending on EV source, recipient cell type, and EV surface composition. In a widely cited mechanistic uptake study, chemical inhibition and knockdown experiments supported predominant roles for clathrin-independent endocytosis and macropinocytosis (including dependence on cholesterol and key GTPases), rather than clathrin heavy chain–dependent pathways, highlighting that EV entry routes are context specific and experimentally testable [115]. For MSC–EVs specifically, functional papers often demonstrate uptake

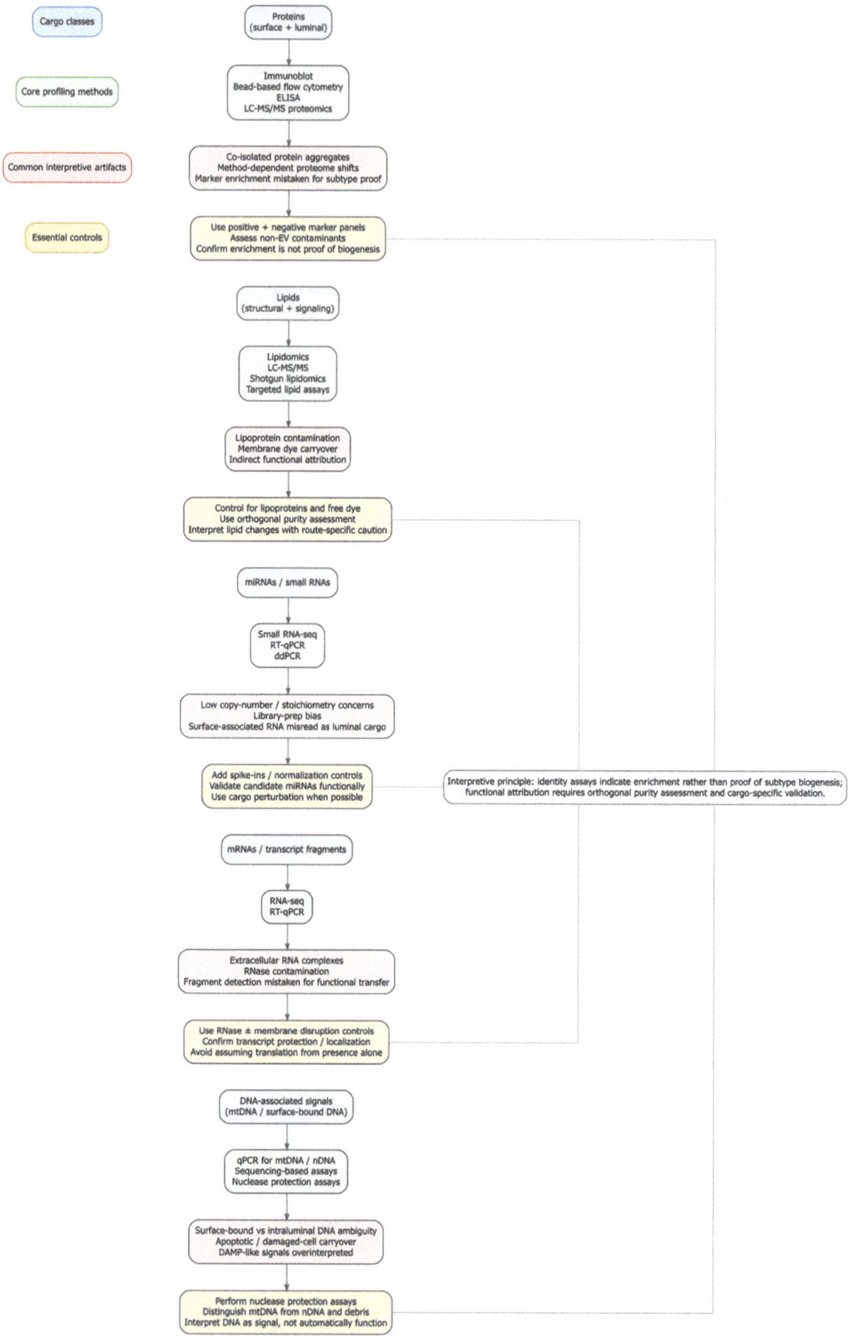

Fig. 4.2 Cargo classes, profiling methods, and interpretive risk controls for MSC–EVs. MSC–EV cargo includes proteins, lipids, RNAs, and DNA-associated signals, but each cargo class requires different analytical methods and different control logic. The figure maps representative cargo classes to core profiling methods, frequent interpretive artifacts, and essential controls needed to support mechanistic claims. Identity assays indicate enrichment rather than proof of subtype biogenesis; therefore, functional attribution requires orthogonal purity assessment and cargo-specific validation

via fluorescent labeling and microscopy, but uptake visualization alone does not define productive cargo release; endosomal escape vs lysosomal degradation remains a central unknown in many systems [112, 123].

4.4.2 *Determinants of Tropism and Targeting*

Tropism determinants can include surface proteins (integrins, tetraspanins, immune checkpoint ligands), glycans, and lipid composition that modulate binding to cell surface receptors and uptake pathways. While much of the integrin-driven "organotropism" evidence comes from tumor EVs and is slightly outside the 2016–2026 window, it remains influential for conceptualizing how EV surface repertoires might encode targeting [125]. MSC–EV-specific targeting determinants are increasingly evaluated via comparative surface protein profiling across different MSC sources, reinforcing that parent cell origin can alter EV surface identity and potentially tropism [126] (Fig. 4.3).

4.4.3 *In Vivo Biodistribution and Pharmacokinetics*

Systemic Delivery: Dominant Liver/Spleen Sequestration with Injury-Enhanced Target Signal In a spinal cord injury model, PET imaging of ^64Cu-labeled MSC–EVs 1 h post-injection showed substantial liver and spleen accumulation, and organ radioactivity analysis at 24 h preserved this trend while revealing limited but significantly higher spinal cord signal in injured versus normal

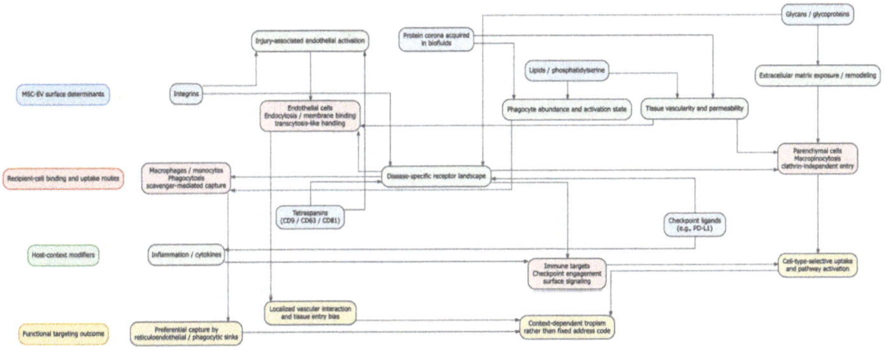

Fig. 4.3 Targeting determinants of MSC–EVs: surface code, protein corona, and recipient cell uptake routes. Tropism of MSC–EVs emerges from EV surface composition and host context rather than from a fixed intrinsic address code. Integrins, tetraspanins, glycans, lipids/phosphatidylserine, checkpoint ligands, and protein corona collectively shape how macrophages, endothelial cells, parenchymal cells, and immune targets bind and internalize EVs. Injury-associated changes in recipient tissues further modify capture and uptake route selection

animals [112]. This pattern aligns with a broader observation: systemic EV dosing often yields high uptake by reticuloendothelial organs (liver/spleen) and lower delivery fractions to distal sites, which has direct implications for dosing strategy (repeat dosing, route selection) and for engineering approaches to evade rapid clearance [112].

Disease Context Can Redirect EV Accumulation and Uptake In a radiation injury bone marrow murine model, DiD-labeled human MSC–EVs injected intravenously preferentially accumulated in liver and spleen with detectable marrow signal; importantly, radiation injury increased MSC–EV uptake in spleen and marrow cells and shifted uptake toward CD11b + and F4/80+ populations. Quantitatively, spleen uptake rose from ~3.57% to ~8.74% and marrow uptake from ~0.75% to ~2.04% after radiation, supporting the idea that injury states can enhance capture/uptake and thereby effective exposure [113] (Table 4.3 and Fig. 4.4).

4.4.4 Critical Appraisal, Gaps, and Future Directions for Uptake/Biodistribution

Labeling Artifacts and Quantitation Limits Lipophilic dyes (DiD/DiR/PKH) can confound biodistribution through dye persistence or labeling of non-vesicular lipid structures; careful dye-only controls and orthogonal labeling (radiolabeling, bioluminescent reporters, magnetic labeling) are increasingly emphasized. Wen et al. explicitly performed DiD-only controls and dose titrations to mitigate free dye interpretations, illustrating best practice in dye-based EV tracking [113].

Table 4.3 Approximate % injected dose per gram (%ID/g) at 24 h after IV injection of ^{64}Cu-labeled MSC–EVs in spinal cord injury vs normal rats

Organ	SCI (%ID/g, approx.)	Normal (%ID/g, approx.)
Liver	4.60	4.36
Spleen	3.92	3.68
Kidney	1.80	2.08
Bowel	0.56	0.60
Blood	0.36	0.24
Lung	0.28	0.20
Heart	0.24	0.20
Stomach	0.08	0.04
Muscle	0.08	0.04
Spinal cord	0.04	0.04

Interpretation: Liver and spleen dominate exposure; kidney provides a secondary sink; other organs show low uptake with modest contextual shifts. Injury-enhanced spinal cord signal is shown separately in the original figure panel and is described as significantly higher in SCI [112]

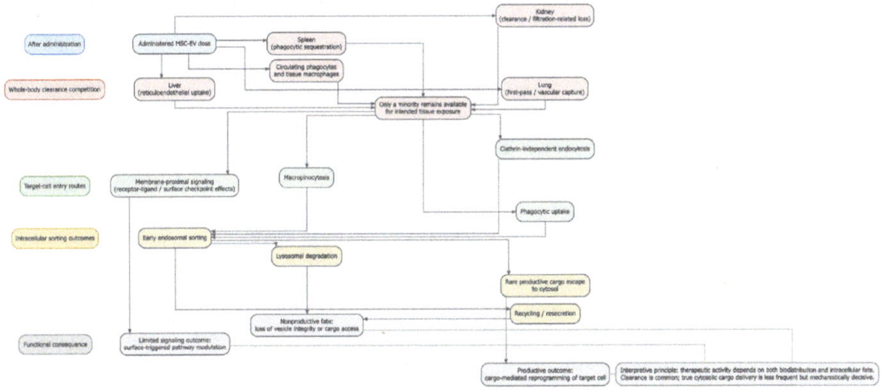

Fig. 4.4 Productive delivery versus clearance of MSC–EVs. After administration, MSC–EVs compete with whole-body clearance sinks including liver, spleen, lung, kidney, and phagocytic cells, so only a minority reaches the intended tissue. At the target cell, therapeutic activity can arise from membrane-proximal signaling or from internalization through macropinocytosis, clathrin-independent endocytosis, and phagocytic uptake. Intracellular sorting then determines outcome: lysosomal degradation and recycling are common nonproductive fates, whereas true cytosolic cargo release is less frequent but critical for cargo-mediated responses

Pharmacokinetics Remains Under-Resolved for MSC Products Many MSC–EV therapeutic studies infer exposure from single timepoint organ imaging, but do not establish full time–activity curves, blood clearance kinetics, or tissue-level effective concentrations. This hinders rational "dose–response in vivo" modeling and comparability across routes (IV vs local vs intranasal) [112, 113].

Future directions include standardized, GMP-compatible labeling for PK studies, linking uptake mechanism perturbation to in vivo distribution shifts, and integrating biodistribution with potency assays to define exposure–response relationships [5, 115].

4.5 Mechanisms of Action Linking MSC–EV Cargo to Functional Outcomes

4.5.1 Regeneration and Tissue Repair Phenotypes

Cardiac Repair with Dose–Response Signals (Moderate Mechanistic Strength) In a rat myocardial infarction model, MSC–EVs injected into infarct border regions improved cardiac function, reduced inflammation and fibrosis, and outperformed MSCs in some measures. The same study demonstrated in vitro dose-dependent enhancement of cardiomyocyte proliferation and dose-dependent reduction of oxidative stress apoptosis, supporting a plausible quantitative relationship

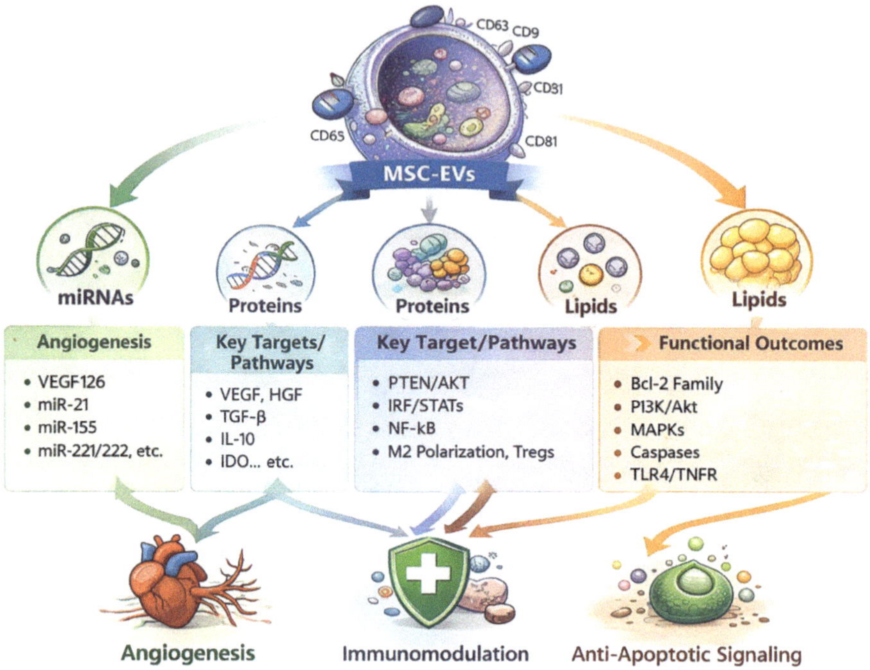

Fig. 4.5 Cargo–function interaction network in MSC–EVs. Integrated schematic linking EV cargo classes (miRNAs, proteins, lipids) to key functional outcomes, including angiogenesis, immunomodulation, and anti-apoptotic signaling pathways

between EV dose and cellular phenotype [123]. Mechanistically, the study used miRNA sequencing (RPKM) to show broadly similar miRNA profiles between MSCs and MSC–EVs, while identifying differentially expressed miRNAs and pathway enrichment (PI3K–AKT and mTOR). This is mechanistically suggestive but not fully causal for a specific miRNA, because individual miRNA perturbations were not the core focus and multiple pathways could mediate observed effects [123] (Fig. 4.5).

Intervertebral Disc Degeneration and Anti-apoptosis via PTEN–PI3K/AKT (High Mechanistic Strength) A disc degeneration model provides one of the more direct causal chains: MSC–EVs were taken up by nucleus pulposus cells, miR-21 was enriched in MSC–EVs and downregulated in apoptotic cells, and exosomal miR-21 alleviated TNF-α-induced apoptosis by targeting PTEN and activating PI3K/AKT signaling; in vivo, intradiscal injection alleviated degeneration and apoptosis markers, and effects were suppressed when exosomal miR-21 was downregulated [114].

4.5.2 Angiogenesis and Vascular Remodeling

Pro-angiogenic activity is frequently attributed to miRNA transfer and downstream modulation of endothelial proliferation, migration, and tube formation assays, but the mechanistic strength varies widely by study. A representative primary study reported that superparamagnetic iron oxide–labeled MSC–EVs improved retinal ischemia and inhibited inflammation by transferring miR-126, consistent with a pro-angiogenic/immunomodulatory miRNA mechanism [127]. Critical caveat: angiogenesis claims are often supported by tube formation and marker expression (VEGF pathway genes), which are necessary but not sufficient; stronger evidence requires (i) miRNA knockdown in parent MSCs to deplete EV cargo, (ii) rescue experiments, and (iii) in vivo perfusion/vascular density measures with adequate blinding and dosing metrics [5].

4.5.3 Immunomodulation and Inflammation Resolution

Immune Checkpoint Mechanisms (High Mechanistic Strength) WJ–MSC-derived sEVs enriched in PD-L1 suppressed CD4+ T-cell activation; PD-L1 neutralization eliminated inhibitory effects, and PD-L1 knockout abrogated both MSC- and sEV-mediated inhibition, linking a specific surface cargo to functional immunosuppression. This study connects EV surface protein cargo to a defined signaling axis (pZAP70–NFAT) downstream of TCR activation [118] (Table 4.4).

Table 4.4 Key MSC–EV studies linking cargo to function

Model system	EV source and type	Dose/route (as reported)	Main findings relevant to MoA	References
Rat myocardial infarction; cardiomyocyte and fibroblast assays	Rat bone marrow MSC–EVs ("MSC-Exo")	20 μg local injection into infarct border; in vitro dose–response assays	Improved LVEF/FS; reduced fibrosis/ inflammation; dose-dependent proliferation and anti-apoptosis; miRNA-seq and pathway enrichment (PI3K–AKT/ mTOR)	[123]
TNF-α-induced nucleus pulposus apoptosis + rat intervertebral disc degeneration	MSC–EVs enriched in miR-21	Intradiscal injection at 1 and 5 weeks post puncture	Exosomal miR-21 targets PTEN → activates PI3K/ AKT → anti-apoptosis; in vivo benefit reduced when exosomal miR-21 downregulated	[114]

(continued)

Table 4.4 (continued)

Model system	EV source and type	Dose/route (as reported)	Main findings relevant to MoA	References
T-cell activation assays; aGVHD patient context	WJ–MSC-derived sEV enriched in PD-L1	10 μg/mL sEV overnight reduced T-cell activation; KO and antibody controls	PD-L1+ sEV required for TCR inhibition via pZAP70–NFAT; genetic knockout and antibody neutralization support causality	[118]
Radiation injury bone marrow murine model; biodistribution	Human MSC–EVs (combined exosome + microvesicle population) labeled with DiD	2 × 10^9 EVs IV tail vein; quantified at 6 h; varied radiation dose	Liver/spleen/marrow accumulation; radiation increases EV uptake; uptake shifts to CD11b+/F4/80+ cells; supports injury-enhanced targeting and cellular sinks	[113]
Spinal cord injury (rat); NET assays; PET biodistribution	Amnion-derived MSC–EVs	100 μg IV 24 h post injury; ^64Cu-labeled EVs for PET; biodistribution at 24 h	Reduced NET formation; mechanistic miR-125a-3p mimic/inhibitor evidence; PET shows liver/spleen dominance with injury-enhanced spinal cord signal	[112]
Oxygen-induced retinopathy model	MSC–EVs (SPIO labeled) carrying miR-126	Route/dose reported in primary paper; cargo emphasized	Angiogenesis/retinal protection linked to miR-126 transfer; inflammation inhibition reported	[127]
Manufacturing-oriented 2D vs 3D culture	Human BM–MSC–EVs	Yield quantified (μg per 1 × 10^5 cells); multiple culture conditions	3D spheroids increase exosome secretion; morphology and spheroid size/density tune output; priming stimuli also assessed	[108]
2D vs 3D MSC culture microenvironments	MSC–EVs with composition profiling	Comparative; EV proteome differences	EV protein composition depends on microenvironment; functional gains not universal → potency must be empirically validated	[110]
Acute lung injury; macrophage polarization	MSC-derived EVs carrying miR-27a-3p	Dose/route in primary paper; mechanistic axis emphasized	Reduced lung injury by promoting M2 polarization and inhibiting NF-κB and inflammasome activation via miR-27a-3p	[124]
MSC–EV fractionation; mitochondrial component analysis	"Ectosomes" (~250 nm) vs "EVs" (~100 nm)	qPCR quantification of mtDNA/nDNA	Ectosomes enriched in mtDNA (~30× vs cells), EVs depleted (3–5×), raising questions about EV-associated mitochondrial components as DAMP-like signals	[122]

Macrophage Polarization and ALI (Moderate-to-High Mechanistic Strength, Limited by Accessibility of Full Methods Here) In acute lung injury, MSC-derived EVs were reported to reduce inflammation and promote M2 macrophage polarization through transfer of miR-27a-3p, with suggested inhibition of NF-κB and inflammasome activation. This provides a mechanistic miRNA hypothesis consistent with broader immunomodulatory models, but strength depends on EV characterization, dosing units, and in vivo controls [124].

Innate Immune Modulation via NET Regulation (High Mechanistic Strength) In spinal cord injury, MSC–EVs reduced neutrophil infiltration and NET formation, and mechanistic experiments showed that the miR-125a-3p mimic reduced NET formation, while the inhibitor reversed exosome effects. This links a specific miRNA cargo to a defined inflammatory effector process that plausibly mediates secondary injury [112].

4.5.4 Dose–Response, Potency Metrics, and Functional Assays

Across models, the field uses heterogeneous potency units: μg EV protein, particle counts, or "EV equivalents" derived from conditioned media volumes. A myocardial infarction study used 20 μg EVs injected locally, while a radiation injury biodistribution study used 2 × 10^9 EV particles and explicitly demonstrated dose-dependent fluorescence signal and uptake-related readouts [113, 123] (Fig. 4.6). This heterogeneity complicates cross-study dose–response interpretation: protein-based dosing can vary with co-isolated proteins, while particle counts shift with detection modality (NTA settings, refractive index assumptions). Current standards urge transparent reporting of quantification methods and multiple orthogonal characterization measures [5].

4.5.5 Critical Appraisal, Controversies, and Future Directions for Mechanisms of Action

Evidence Strength Is Uneven Across Endpoints The strongest mechanistic studies (PD-L1 checkpoint, miR-21–PTEN axis, miR-125a-3p NET suppression) share features: defined EV identity assays, explicit perturbations that alter EV cargo (knockout or miRNA manipulation), and functional reversal/rescue experiments [118].

In contrast, many "repair" claims rest on broad profiling and pathway enrichment, which is mechanistically plausible but can be confounded by isolation method–dependent cargo differences and mixed-vesicle preparations [5, 100].

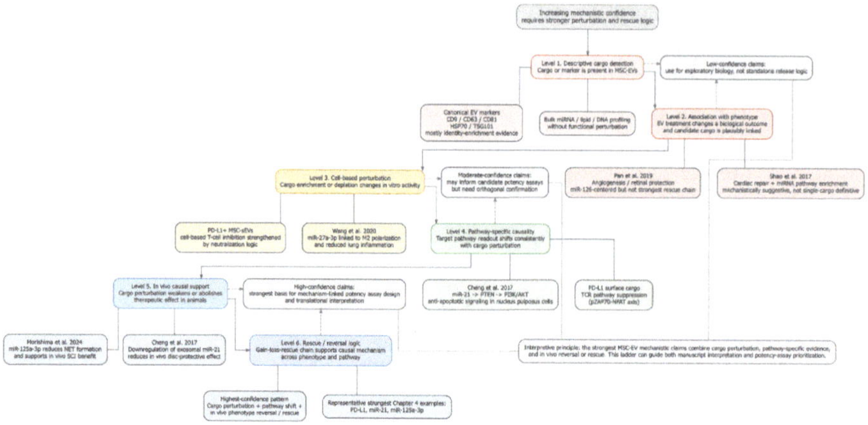

Fig. 4.6 Mechanistic confidence ladder linking MSC–EV cargo to biological outcomes. Mechanistic claims in MSC–EV studies vary in evidentiary strength. This figure organizes representative Chap. 4 examples on a ladder from descriptive cargo detection to causal in vivo reversal, highlighting how surface proteins such as PD-L1 and miRNA cargos such as miR-21 and miR-125a-3p reach stronger mechanistic confidence when perturbation and rescue experiments are included. The framework can guide potency assay design and interpretation of translational claims

Stoichiometry Remains a Core Controversy Even when miRNAs are clearly present and functional in perturbation studies, open questions remain about how many miRNA copies per vesicle, what fraction of vesicles carry the active moiety, and whether observed effects reflect a small high-potency subpopulation. This problem is amplified by the heterogeneity documented in EV subpopulations and by isolation method effects on composition [100, 111, 119].

Translational Bottleneck: Linking Exposure to Response Biodistribution data show major systemic sequestration in liver/spleen, while therapeutic claims often presume sufficient target–tissue delivery. Without PK modeling and tissue exposure metrics, it is difficult to reconcile low target site fractions with robust biological effects, suggesting that (i) small delivered doses may be sufficient in highly sensitive inflammatory cascades, (ii) immune organ modulation could indirectly mediate distal benefit, or (iii) labeling methods over- or underestimate true delivery [112, 113, 115].

Future research priorities include (1) potency units tied to mechanism; (2) single-EV multi-omics to resolve functional subpopulations; (3) standardized, orthogonal biodistribution methods (PET + mass balance + tissue microscopy); and (4) manufacturing controls showing cargo stability across donors, passages, preconditioning, and storage [5, 100].

Chapter 5
Isolation, Characterization, Tracking, and Stability of MSC–EVs

Abstract This chapter addresses the isolation, characterization, tracking, and stability of MSC-derived extracellular vesicles. It compares major isolation techniques and emphasizes the principle that method selection defines product composition. The chapter discusses MISEV-compliant characterization strategies, potency testing frameworks, and critical quality attributes (CQAs). It also evaluates storage conditions, stability challenges, and strategies to preserve EV bioactivity for clinical applications.

Keywords EV isolation · Characterization · Quality control · Potency assays · Stability

MSC–EVs are increasingly treated as biological drug products rather than generic "nanoparticles." Their apparent simplicity (acellular, non-replicative) conceals a set of interlinked technical realities: (i) EV preparations are inherently heterogeneous and co-isolate with non-vesicular extracellular particles (NVEPs) and soluble macromolecules; (ii) the isolation method defines the product (composition, impurities, and potentially mechanism-linked biological activity); and (iii) clinically credible translation requires MISEV-compliant characterization, a fit-for-purpose potency strategy, and stability-informed formulation/storage. These principles are formalized in the MISEV2018 and MISEV2023 consensus statements and should guide not only reporting but also manufacturing and quality control design [5, 71].

For downstream processing, there is no universally "best" isolation platform. MISEV2018 explicitly frames methods along a recovery vs specificity spectrum (from "high recovery/low specificity" to "low recovery/high specificity") and stresses that method choice must be driven by the scientific or clinical question [5]. In the last decade, scalable, chromatography- and filtration-centered workflows have matured as credible paths toward GMP-compatible manufacture, while ultracentrifugation remains common for discovery-scale work but is limited by throughput, operator dependence, and potential EV damage/aggregation [96, 128, 129].

Functionally, potency and quality cannot be inferred from particle counts alone. Regulatory guidance for potency in complex biologics emphasizes that potency assays should measure a defined biological effect linked as closely as possible to

N. M. Mussin, A. Tamadon, *Mesenchymal Stem Cells and Their Extracellular Vesicles*, https://doi.org/10.1007/978-3-032-36437-1_5

mechanism(s) of action and that validated in vitro assays are preferred for routine batch release when feasible [130]. For MSC–EVs, where mechanisms are often multifactorial and indication specific, a pragmatic, regulator-aligned approach is a potency assay matrix combining orthogonal biochemical and cell-based readouts to monitor consistency and clinically meaningful variability [131, 132].

Finally, storage and handling are not "afterthoughts." Multiple primary studies show that storage temperature and formulation measurably affect MSC–EV bioactivity. For example, bone marrow MSC–EV pro-angiogenic and anti-inflammatory activity can persist for weeks at −20 °C/−80 °C and after lyophilization, while activity can decline at 4 °C or room temperature depending on duration and readout [133]. Cryo−/lyoprotectants such as trehalose can reduce aggregation and preserve functional activity in several EV systems, supporting rational formulation design rather than default PBS-only storage [134–136].

5.1 Isolation Methods

Isolation of MSC–EVs is best understood as product definition rather than a purely preparative step (Table 5.1). Two factors dominate method selection (Fig. 5.1): (i) the intended use (exploratory biology vs quantitative analytics vs therapeutic manufacture) and (ii) the contaminants most likely to confound interpretation or safety (soluble proteins, lipoproteins, extracellular ribonucleoproteins, polymer carryover, endotoxin, adventitious agents). MISEV2018 explicitly states that complete "isolation" is often unrealistic and that separation/enrichment should be selected and reported in a way that matches the downstream question [5].

A second foundational constraint is upstream variability. In this chapter, the MSC tissue source is unspecified and should be treated as variable. Tissue source and culture conditions (media additives, priming, seeding density, passage) measurably alter EV yield and cargo and can shift functional readouts, so isolation protocols must be embedded in a broader process definition [131, 137].

MISEV2018 provides a highly practical grid of separation approaches by expected recovery and specificity, emphasizing that precipitation and basic filtration can recover large amounts of extracellular material but with lower EV specificity, while density gradients and affinity methods can enrich specific EV subtypes but often at the cost of yield and scalability [5].

5.1.1 Method-by-Method Detail and Recommended Protocol Elements

Differential Ultracentrifugation (dUC) dUC remains a common baseline method, but MISEV emphasizes that centrifugation must be reported with sufficient detail to be reproducible (g-force, rotor type, k-factor, time, temperature, brake set-

Table 5.1 Comparative overview of MSC–EV isolation and purification options

Method (as requested)	Core principle	Typical recovery vs specificity framing	Major strengths	Key limitations/ artifacts	Scalability and automation	Indicative cost profile
Differential ultracentrifugation (dUC)	Sequential pelleting by size/density proxy	Often intermediate recovery/ intermediate specificity; highly context dependent	Widely used; low consumables; compatible with many labs	Co-pelleting of protein aggregates/ NVEPs; aggregation; operator/rotor dependence; limited throughput	Limited for large volumes (batch, labor intensive)	High CAPEX (ultracentrifuge), low per-run consumables
Density gradients (sucrose/iodixanol)	Separation by buoyant density	Lower recovery/ higher specificity	Improves purity; can resolve EV subpopulations and co-isolates	Time-consuming; low throughput; may require fractionation expertise	Poor for high-volume clinical manufacturing (unless automated chromatography-like platforms are used)	Moderate; labor and analytical burden high
Size exclusion chromatography (SEC)	Size-based elution through porous matrix	Intermediate recovery/ intermediate-to-high specificity (for soluble proteins)	Gentle; reduces soluble protein carryover; preserves function in comparative studies	Requires fraction collection; can co-elute some particles (matrix and sample dependent)	Scalable and automatable (FPLC/ columns)	Moderate-to-high consumables; lower labor with automation
Ultrafiltration (incl. TFF)	Membrane-based concentration/ size cutoff	High recovery/ low-to-intermediate specificity (alone)	Processes large volumes; fast; scalable; widely used upstream of chromatography	Membrane adsorption; shear; concentrates contaminants without additional purification	High (TFF is manufacturing standard)	Moderate equipment; moderate consumables

(continued)

Table 5.1 (continued)

Method (as requested)	Core principle	Typical recovery vs specificity framing	Major strengths	Key limitations/ artifacts	Scalability and automation	Indicative cost profile
Precipitation (PEG, charge based, kits)	Polymer crowding/ charge-driven precipitation	High recovery/low specificity	Easy; low equipment; handles many samples	Co-precipitates proteins/polymers; purity variability; added reagent impurities require controls	Moderate (but purification steps often still needed)	Low equipment; reagent cost varies (kits often high)
Microfluidics	On-chip size/ charge/acoustic/ affinity capture	Variable; often higher specificity for small volumes	Integrates isolation + detection; low sample volume; rapid	Platform-dependent performance; standardization lacking; not yet proven for manufacturing scale	Low for manufacturing (today); promising for analytics	High development/ device cost; low per-test volume
Immunoaffinity (beads/resins/ aptamers)	Capture via surface ligands	Lower recovery/ higher specificity (subset focused)	Enables subpopulation enrichment and mechanistic/ biomarker specificity	Marker bias (heterogeneous marker distribution); elution challenges; expensive	Low to moderate; column resins emerging but still niche	High reagents; moderate equipment

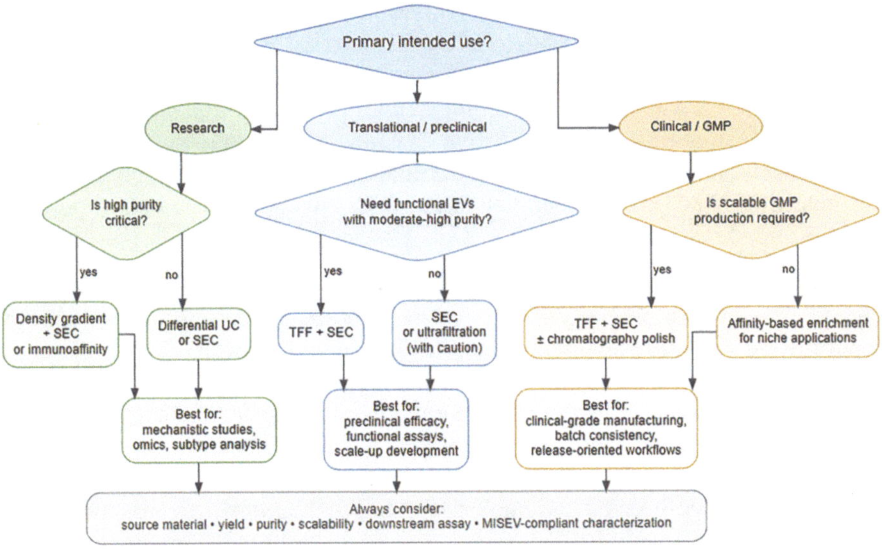

Fig. 5.1 Decision framework for selecting EV isolation methods. Workflow guiding method selection based on intended use (research vs clinical), required purity, scalability, and downstream application constraints

tings) and that low-specificity recovery steps should be complemented by characterization and, when needed, secondary purification [5]. In functional comparisons, SEC-derived EV preparations have sometimes shown higher activity than UC-derived preparations, supporting the concern that pelleting can alter EV state or co-isolate inhibitors [5].

Density Gradients Density gradients are central to resolving EV heterogeneity and improving purity. Primary proteomic work demonstrates that "small EV" fractions can include multiple subtypes and co-isolated particles and that immunoisolation and density-based separation reveal meaningful subpopulation differences [111]. This directly impacts MSC–EV interpretation: a "therapeutic EV" effect may be driven by a subset enriched or depleted by gradients [111].

Size Exclusion Chromatography (SEC) and Chromatography Hybrids SEC is increasingly used for therapeutics because it is comparatively gentle and effective at depleting soluble proteins. Multiple primary studies support SEC-centered workflows: (i) SEC–EVs can retain higher functional signaling readouts than UC–EVs in head-to-head comparisons [5]; (ii) UF–SEC provides EV isolates suitable for proteomics and functional work with improved separation from free proteins [138]; (iii) bind–elute SEC can achieve high recoveries (~80%) with reproducibility and scalability [128]; and (iv) automated/fast implementations report high yields and short run times, addressing a core bottleneck for scale [139].

Ultrafiltration (Spin Filters, Ultrafiltration Concentrators, and TFF) Ultrafiltration is best viewed as a concentration step, not a full purification step, because it can concentrate both EVs and non-EV macromolecules. TFF has strong evidence for yield, reproducibility, time, cost, and scalability advantages over ultracentrifugation in comparative isolation studies [107]. Recent MSC-focused work and reviews argue that ultrafiltration can be viable for both lab-scale and industrial MSC–sEV preparation when paired with complementary purification [140].

Precipitation (PEG and Charge Based) Precipitation methods are operationally attractive but are generally challenged by co-precipitation of proteins and other impurities. MISEV2018 explicitly cautions that some protocols—especially commercial kits—may introduce polymers or beads/antibodies that affect profiling and functional studies and may render preparations unusable for therapeutic applications without additional controls and purification [5]. A key mitigation is secondary cleanup: PEG precipitation followed by washing and ultracentrifugation can remove substantial PEG-co-precipitated molecules, improving downstream compatibility [141].

Microfluidics Microfluidic EV isolation platforms offer rapid processing and integration with detection, with methods spanning filtration, inertial focusing, viscoelastic flows, acoustic approaches, and on-chip immunoaffinity capture. High-quality reviews emphasize promise for analytics and point-of-care use but highlight that standard performance characterization and manufacturing-scale feasibility remain limitations [142].

Immunoaffinity Affinity isolation is best considered subpopulation enrichment and is heavily influenced by marker distribution across EVs. Primary and translationally oriented work shows emerging formats beyond beads, but these remain more specialized than SEC/TFF for broad manufacturing [126]. MISEV2023 reinforces that labeling and marker-based methods require careful reporting and controls because they can bias which EV subsets are detected or isolated [71].

Because the MSC source, culture system, and intended indication are unspecified, the protocols below are framed as recommended elements and parameter classes rather than single fixed SOPs. Method parameters should be tuned using orthogonal characterization and, for clinical products, potency-linked comparability testing [5, 131] (Table 5.2).

Recent contributions from the authors' group have compared EV isolation techniques for clinical applications and discussed ligand-based affinity purification using LEAP column chromatography [143, 144]. Affinity-based approaches may improve enrichment of selected EV populations, but their clinical utility depends on ligand specificity, recovery efficiency, elution conditions, scalability, residual–reagent control, cost, and demonstration that vesicle integrity and biological activity are preserved. Such products should therefore be described as affinity-enriched EV subsets rather than universally representative MSC–EV preparations.

Table 5.2 Fit-for-purpose recommended process "recipes" (conceptual protocols)

Intended use	Recommended isolation/purification sequence	Rationale and key controls
Discovery functional studies (moderate purity)	Pre-clear conditioned media → dUC (high-speed pelleting) → wash/resuspend → optional SEC cleanup	Common baseline; requires strong impurity controls and EV-associated function controls per MISEV. SEC can reduce soluble protein confounding and may preserve function [5]
Omics/mechanistic studies (high purity)	Pre-clear → concentration (TFF or spin UF) → SEC fractionation → optional density gradient or AEX polish	UF–SEC improves separation from free proteins. AEX has scalable potential for MSC-derived feedstocks. Use MISEV marker categories to quantify EV features and co-isolates [138]
Clinical/GMP-oriented manufacture (closed, scalable)	Defined upstream (xeno-free or EV-depleted supplements) → continuous harvest/clarification → TFF concentration → SEC and/or chromatography (AEX/BE-SEC) → sterile filtration → fill–finish	Scalable frameworks demonstrated for large volumes and GMP-compatible workflows. Sterility/endotoxin controls and validated potency release criteria required [145]
Subpopulation/marker-specific analytics	Pre-clear → SEC (reduce soluble background) → immunoaffinity capture (bead/resin) → gentle elution (validated)	Enriches defined subsets but introduces marker bias; must be interpreted in the context of EV heterogeneity [111]

5.2 Characterization per MISEV2018 and MISEV2023

MISEV2018 and MISEV2023 establish the field's consensus that EV characterization requires (i) quantitative description of the source material and EV preparation, (ii) demonstration of EV features using a structured marker framework, (iii) assessment of likely co-isolates/impurities, and (iv) fit-for-purpose functional validation with appropriate controls [5, 71] (Table 5.3 and Fig. 5.2). For MSC–EVs, these principles also address a translational concern: many reported "MSC–EV effects" vary across studies because MSCs and their EVs are responsive to microenvironment and manufacturing conditions, so standardized, comparable characterization is mandatory for credible potency and stability claims [131].

MISEV2018's "steps of EV characterization" table provides operational minimal expectations: describe source quantitatively; provide global quantification (commonly particle number and total protein) and consider ratios as purity metrics; demonstrate at least three positive markers spanning membrane-associated and cytosolic categories and at least one negative/depleted marker; and report single-vesicle/structural analysis using complementary methods where feasible [5]. MISEV2023 retains and reinforces the five-component protein framework for EV preparations and adds explicit emphasis on NVEP co-isolates and on technique-specific reporting [5].

Table 5.3 Practical MISEV-compliant characterization checklist for MSC–EVs

Characterization domain	Minimal (MISEV aligned)	Recommended for robust/clinical translation
Source reporting (MSC + harvest)	Report MSC source as specified/ unspecified; report culture conditions, harvest timing, and starting volume/cell equivalents	Track donor/tissue metadata, passage, seeding density, priming, medium supplement EV-depletion method, and harvest frequency (critical for comparability)
Particle size distribution	One particle sizing method appropriate for size range	Orthogonal sizing and technique-specific reporting to detect biases and aggregates
Particle concentration/dose metric	Report at least one global metric (often particle count and/or total protein)	Dual metrics (particle count + protein + ratios) and indication-relevant dose normalization
Morphology/ membrane-limited structure	EM-based imaging to show vesicular morphology (where feasible)	Cryo-EM or careful negative-stain EM with reporting of preparation artifacts; quantify morphology on representative fields
Protein markers (identity + purity controls)	Analyze ≥1 marker from categories 1–2 and ≥1 marker from category 3 (purity) using the MISEV marker framework	Apply the five-component framework and show enrichment/depletion relative to source; include a panel appropriate for MSC origin and impurities
RNA content	Report RNA only as supportive information (not a purity surrogate)	Use spike-ins and controls for co-isolated RNA complexes; treat total RNA as secondary unless linked to function
Functional assay linkage	Demonstrate EV-associated activity with appropriate controls (EV-depleted fractions, process controls)	Define potency-linked assays for intended indication; add specificity controls, dose–response, and stability-indicating functional readouts
In vivo EV studies (if performed)	Report labeling and dosing parameters	Control for label effects on PK/BD; include ex vivo quantification and label-only controls (MISEV2023)

5.2.1 *Protein Marker Framework (What to Measure and Why)*

Both MISEV2018 and MISEV2023 advise that EVs lack universal subtype-specific markers; therefore, characterization should use category-based evidence. The canonical, widely used EV-associated markers remain tetraspanins (CD9/CD63/CD81) and ESCRT-related proteins (TSG101/ALIX), but these are not strictly exclusive to EVs; thus, purity controls and depleted markers are essential [5, 71]. MISEV2023 explicitly organizes proteins into five categories: (1) transmembrane/GPI-anchored EV-associated proteins, (2) cytosolic proteins expected in EVs, (3) major non-EV co-isolated components, (4) intracellular compartment proteins (context dependent), and (5) secreted proteins that co-recover with EVs including corona components; it explicitly notes that at least one protein each from categories 1–3 should be analyzed as EV hallmarks and for co-isolate assessment [71].

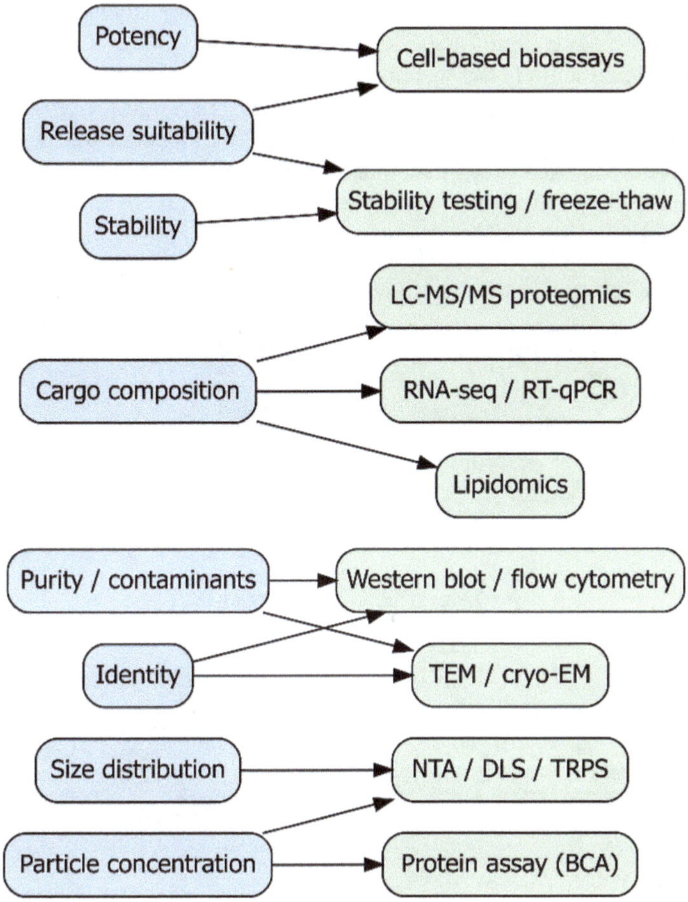

Fig. 5.2 Analytical toolbox mapped to critical quality attributes (CQAs) for MSC–EV therapeutics. Different analytical methods interrogate distinct CQAs required for EV product definition and release. Particle sizing techniques (NTA/DLS/TRPS) assess size and count; imaging and marker assays evaluate identity and purity; omics approaches define cargo composition; functional bioassays assess potency; and stability studies determine storage robustness. No single assay captures all CQAs, necessitating a multi-assay matrix aligned to mechanism of action and regulatory expectations

5.2.2 Particle Quantification and Morphology: Practical Cautions

MISEV2018 underscores that particle counting methods can overestimate EV number because they detect co-isolated particles (lipoproteins, protein aggregates) and that total protein can overestimate due to contaminant proteins, especially in lower-specificity separations [5]. Thus, particle count, protein mass, and ratios should be interpreted in the context of the separation method and impurity markers rather than as absolute "dose equivalents" [5].

Single-particle methods are increasingly used to bridge bulk and single-vesicle characterization and can support subpopulation and marker co-localization claims when appropriately controlled [146]. For MSC–EV therapeutics, these methods are most valuable when used to validate that a manufacturing process yields consistent marker-bearing subpopulations (identity) and to detect shifts in co-isolated particles or aggregation (quality) [146].

5.3 Potency Testing and Quality Control

Potency and QC are where MSC–EV development transitions from "EV research" to "biopharmaceutical development" (Figs. 5.3 and 5.4). The EMA guidance for cell-based immunotherapy products—while not EV specific—provides a

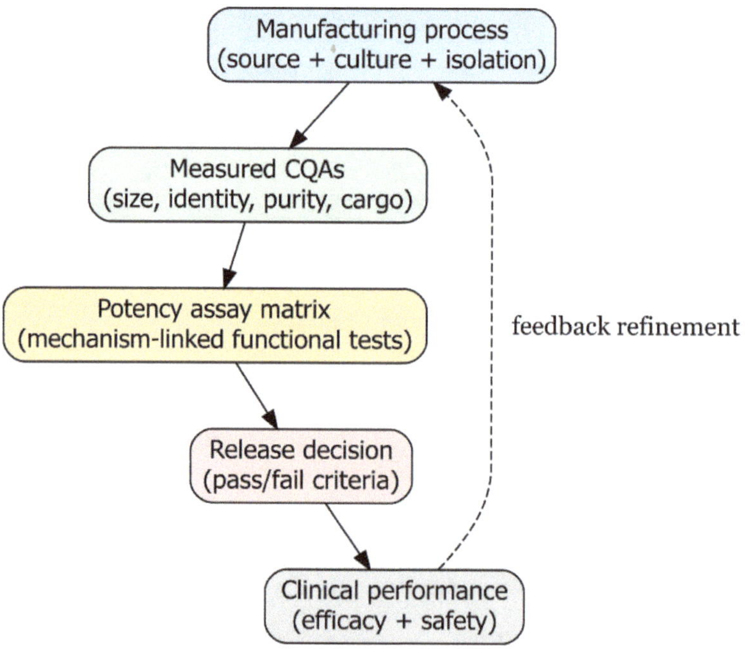

Fig. 5.3 CQA–potency–release integration framework for MSC–EV product development. MSC–EV therapeutic performance emerges from a hierarchical relationship between manufacturing processes, measurable CQAs, and mechanism-linked potency assays. CQAs define product identity and consistency, while potency assays translate these attributes into biological activity. Release decisions depend on integrated evaluation of CQAs and potency, and clinical outcomes feed back to refine manufacturing and assay design, forming a continuous development loop aligned with regulatory expectations

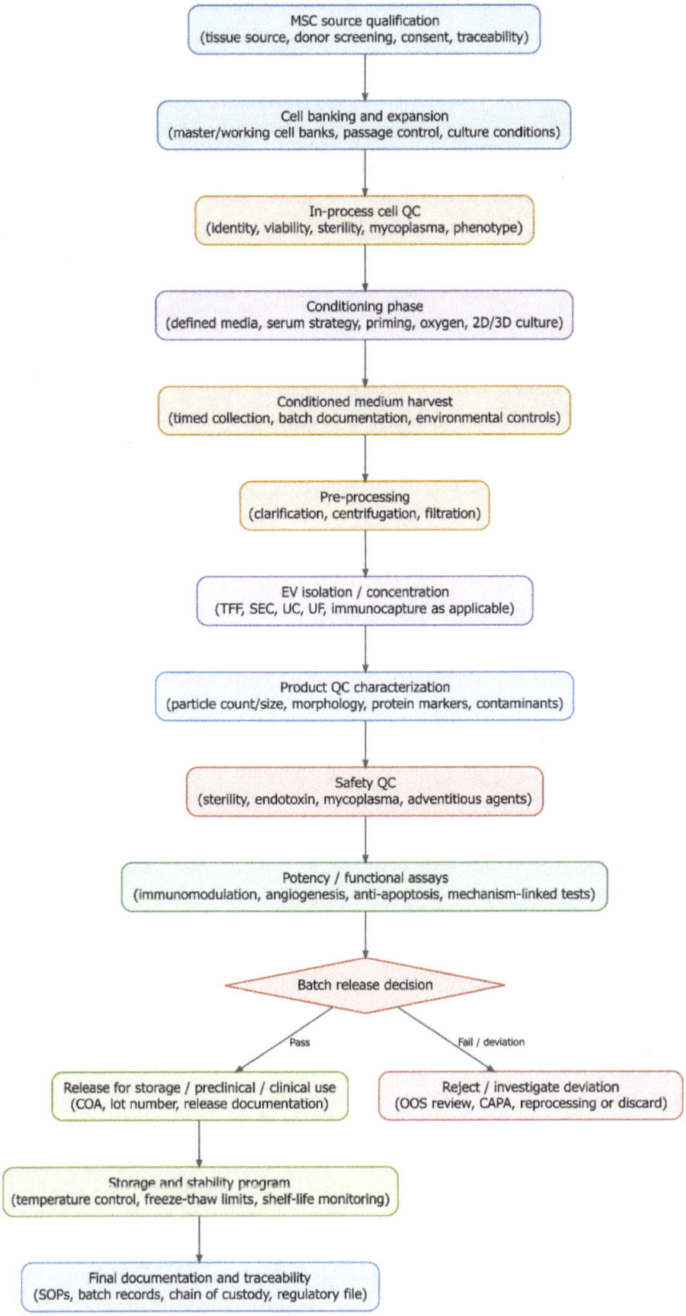

Fig. 5.4 Proposed manufacturing quality control (QC) workflow for Mesenchymal Stromal/ Stem Cell (MSC)–derived Extracellular Vesicle (EV) products. The flowchart summarizes key stages from donor and tissue source qualification through cell banking, expansion, conditioning,

(continued)

regulator-relevant template: potency is a quantitative measure of biological activity; assays should be based on a defined biological effect as close as possible to mechanism(s) of action/clinical response; and in vitro assays are preferred for routine batch release when they can monitor product consistency and correlate with intended effect [130]. This logic aligns closely with MSC–EV translation needs, even though EVs are nonliving: EV products are complex, multi-component biologics with difficult-to-attribute activity and significant batch variability risk [130].

High-quality MSC–EV potency literature converges on a pragmatic approach: because MSC–EV mechanisms are often complex and indication specific, a single potency assay is rarely sufficient. Instead, potency testing should be designed as an assay matrix combining orthogonal biochemical and cell-based assays that quantify EV attributes relevant to intended biological activity, while remaining feasible within shelf-life constraints and reagent availability [131]. Importantly, Gimona et al. emphasize that different manufacturing processes can generate different MSC–sEV products with qualitative/quantitative differences in identity and potency; therefore, potency criteria must follow (and lock) the manufacturing definition [131].

A concrete example of a release testing philosophy is provided by a clinical testing manufacturing framework for umbilical cord MSC–EVs that specifies parental cell identity/viability, EV quantity and size (NTA), EV marker identity, microbial impurities (endotoxin/sterility/mycoplasma), and additional informative tests (RNA profiles, cytokine profile, functional assays such as T-cell inhibition or neuroprotective activity) [132]. The specific numeric thresholds in that report should not be treated as universal; rather, the framework illustrates how to translate EV characterization into release criteria and critical quality attributes (CQAs) [132].

5.3.1 Critical Quality Attributes and Suggested Release Criteria Domains (MSC–EV Therapeutics)

For MSC–EVs intended for human use, CQAs and release criteria typically cluster into identity, purity/impurities, quantity/dose, biological activity/potency, and safety/microbiology, with stability indicating assays spanning multiple categories. Rohde et al. provide an illustrative CQA grid including parental cell marker profile, EV concentration and size by NTA, EV marker identity (including CD63/CD81/

Fig. 5.4 (continued) conditioned-medium harvest, pre-processing, EV isolation, and batch-level product characterization. Critical QC checkpoints include in-process cell testing, physicochemical EV characterization, safety testing, and potency assays linked to the intended mechanism of action. Batch release is based on predefined acceptance criteria, after which released material proceeds to storage, stability monitoring, and full documentation for traceability. Products that fail specifications require deviation investigation and corrective actions. This workflow is intended as a conceptual manufacturing framework for reproducible and translationally relevant MSC–EV production

TSG101 panels), endotoxin limits, sterility, and mycoplasma negativity, alongside informative functional readouts and RNA/cytokine profiling [132].

Gimona et al. [131] further highlight practical constraints that shape assay selection for release: assays must be quantitative and robust, but also rapid enough not to consume product shelf-life; and they must rely on reagents/cell systems with lot-to-lot consistency, suggesting the use of banked cell lines and standardized reagent qualification for potency assays.

5.3.2 *Batch-to-Batch Variability Drivers and Manufacturing Comparability*

Batch variability in MSC–EVs arises from (i) MSC source and donor variability, (ii) culture condition sensitivity (media supplements, priming, oxygen), (iii) harvest/conditioning schedules, (iv) downstream separation method effects, and (v) storage/handling. Reviews and consensus work emphasize that "MSC–EV" is not a single entity: MSCs release EV subpopulations with distinct molecular properties, and separation methods can enrich different subsets, complicating comparability and potency consistency [5, 111, 147]. Consequently, manufacturing changes (scale-up, new filters/columns, new media) should trigger "comparability" assessments anchored in CQAs and potency-linked readouts, not only particle count [147].

5.3.3 *Regulatory Considerations and Safety Signals Relevant to MSC–EV Translation*

In the United States, FDA communications make two points directly relevant to MSC–EV developers: (i) there are no FDA-approved exosome products and exosome products intended to treat disease generally require FDA approval and (ii) serious adverse events have been reported from unapproved products marketed as containing EVs, prompting public safety alerts and enforcement actions [148]. A state health advisory describing Nebraska-associated adverse events and referencing FDA and CDC warnings underscores the real-world consequences of inadequate manufacturing controls and regulatory noncompliance [148].

From a development perspective, these safety communications reinforce why potency, purity, and microbial safety testing must be embedded in a GMP-capable quality system rather than treated as optional "research characterization." Recent regulatory scholarship highlights the need for clear, enforceable frameworks as regenerative medicine and EV-like products proliferate outside formal approval pathways [149].

5.4 Labeling and Tracking of EVs

EV tracking serves three distinct scientific needs (Fig. 5.5): (i) pharmacokinetics and biodistribution (where do administered EVs go, and how long do they persist?); (ii) cellular uptake and trafficking (which cells internalize EVs, and do cargos reach cytosol?); and (iii) mechanism-of-action support (does biodistribution align with therapeutic effect?). MISEV2023 explicitly recommends reporting all details of labeling/detection technologies, reporting administration parameters for exogenous EV studies, and controlling for label effects on biodistribution, PK, and function. It also warns that a signal can represent labeled non-EV substances (free dye, detached label) and that different behaviors of endogenous vs exogenous EVs should be considered [71].

Tracking methods trade off sensitivity, quantifiability, spatial resolution, and artifact risk. Broadly: nuclear imaging is highly sensitive and quantifiable but requires radiochemistry; optical fluorescence is convenient but artifact-prone; bioluminescence has low background but limited spatial resolution; and MRI methods provide anatomical context but often lower sensitivity and risk confounding from label transfer. These trade-offs should be explicitly linked to the study question rather than chosen by convenience alone [71].

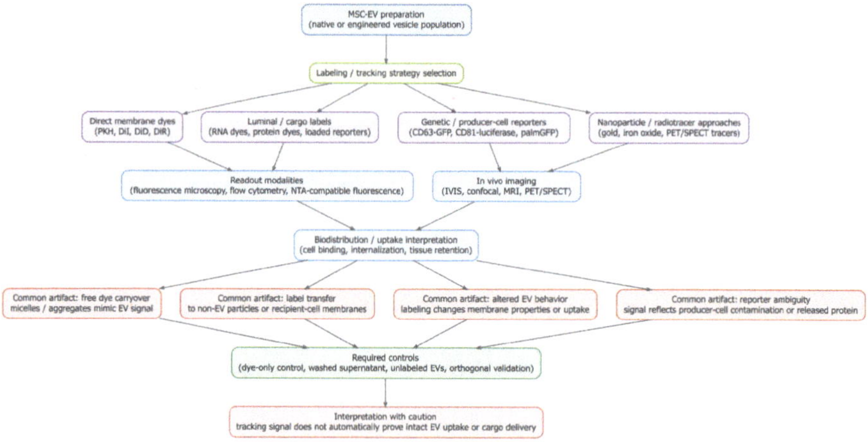

Fig. 5.5 Schematic overview of common labeling and tracking approaches used for mesenchymal stromal/stem cell–derived extracellular vesicles (MSC–EVs), together with major sources of experimental artifact. Labeling strategies include direct membrane dyes, luminal or cargo-associated labels, genetic reporter systems introduced into producer cells, and nanoparticle- or radiotracer-based methods. These approaches are typically read out by fluorescence imaging, flow cytometry, nanoparticle tracking analysis, or in vivo imaging platforms such as IVIS, MRI, and PET/SPECT. However, interpretation is complicated by several well-recognized artifacts, including free dye carryover, micelle or aggregate formation, label transfer to non-EV structures or recipient cell membranes, reporter ambiguity, and possible alteration of EV physicochemical behavior after labeling. The figure emphasizes that tracking signal alone does not prove intact vesicle uptake or functional cargo delivery and that rigorous controls and orthogonal validation strategies are required

5.4.1 Practical Comparison of Tracking Modalities

Fluorescent Dyes (Lipophilic Membrane Dyes, NIR Dyes) Lipophilic dyes remain widely used for EV uptake and biodistribution studies due to simplicity, but they can generate misleading signals via dye aggregates, dye transfer to non-EV membranes, and persistence of free dye beyond EV clearance. MISEV2018 already flags that dyes and antibodies can form particles and that such labeling is a confounder for EV-specific measurements; MISEV2023 reiterates that labeling can alter biodistribution and detectability thresholds, requiring controls. Comparative imaging work in MSC–EV contexts has contrasted PKH-based labels with alternative labeling strategies and highlights the need for label-only and process-matched controls [5, 71].

Genetic Reporters Genetic labeling supports tracking of endogenously produced EVs and can reduce free dye artifacts, but it introduces its own caveats: overexpression can perturb EV biology, labels may tag only certain EV subtypes, and fluorescent proteins can be quenched in acidic compartments; MISEV2023 explicitly discusses such caveats and suggests knock-in strategies and/or multiple EV markers as mitigation. Reporter system reviews provide frameworks for interpreting transfer and uptake assays, emphasizing that observed reporter signal may not equate to functional cargo delivery [71].

Radiolabeling (SPECT/PET) Radiolabeling offers high sensitivity and whole-body quantification. A representative primary example is ^111In labeling of MSC–EVs followed by μSPECT/CT imaging to track in vivo fate, addressing the long-standing limitation that optical dyes can be difficult to quantify in deep tissues [150]. Such studies typically show rapid uptake in clearance organs (often liver/spleen) after systemic delivery; however, label stability and release must be validated [150].

MRI Contrast Agents (SPIO/USPIO) MRI-based tracking provides anatomical resolution but is generally less sensitive than nuclear imaging, and free nanoparticle uptake by the reticuloendothelial system can mimic EV biodistribution. In MSC–EV imaging studies, SPIO-based labeling has been explored alongside fluorescent labels to compare signal specificity and cellular uptake, underscoring that multimodal labeling can aid interpretation but does not eliminate confounders [151].

5.4.2 Biodistribution Examples and Interpretation Principles

Biodistribution can be state dependent. For instance, radiation exposure can alter MSC–EV biodistribution and biological effects, illustrating that disease context and host physiology shape apparent targeting and clearance [113]. Therefore, "biodistribution" should be treated as a CQA adjacent attribute for therapeutics: route, dose, labeling strategy, and disease state can shift both distribution and functional impact, and MISEV2023 urges careful reporting of these parameters [152].

5.5 Storage, Freezing, Stability, and Shelf-Life

MISEV2018 highlights that storage and retrieval conditions of both the matrix (conditioned medium, biofluid) and isolated EVs can affect EV number, aggregation, and function, and it explicitly prompts authors to report buffers, freezing/thawing, lyophilization, cryoprotectants, and freeze–thaw cycles because these variables can change EV activity [5]. For MSC–EV therapeutics, storage is not merely logistical: it determines whether potency-linked CQAs stay within specification through release, shipping, and administration [5].

5.5.1 What Stability Data Show in MSC–EV Studies (Themes Supported by Primary Datasets)

A major primary contribution is a systematic evaluation of storage temperature and duration on bone marrow MSC–EVs, assessing both physical properties and functional bioactivity (Fig. 5.6 and Table 5.4). In that work, MSC–EVs retained key pro-vascularization and anti-inflammatory bioactivity for up to ~4 weeks after storage at −20 °C and −80 °C and after lyophilization, while activity declined at room temperature and/or 4 °C depending on assay and duration [133]. The same study also evaluated cargo-loaded EVs and showed that cargo-associated functional activity can persist even when measured cargo levels decrease, emphasizing that stability must be defined in terms of functional output, not only cargo quantification [133].

A systematic review by Ahmadian et al. [153] demonstrated that EV preservation outcomes are influenced by storage temperature and duration, buffer composition, container characteristics, freezing rate, thawing procedure, and freeze–thaw history. Across the reviewed studies, repeated freeze–thaw cycles were frequently associated with particle loss, aggregation, and deterioration of RNA, protein, or functional readouts, whereas deep-freeze storage and appropriately designed stabilizing formulations generally provided better preservation. These findings support aliquoting EV preparations, minimizing freeze–thaw exposure, documenting the complete storage history, and establishing shelf-life using both physicochemical measurements and potency-linked functional assays rather than particle count alone.

Cryopreservation strategies increasingly incorporate disaccharides. Trehalose has primary evidence as a stabilizer that can reduce aggregation and support EV integrity through freeze–thaw stress, and trehalose-based cryopreservation has been

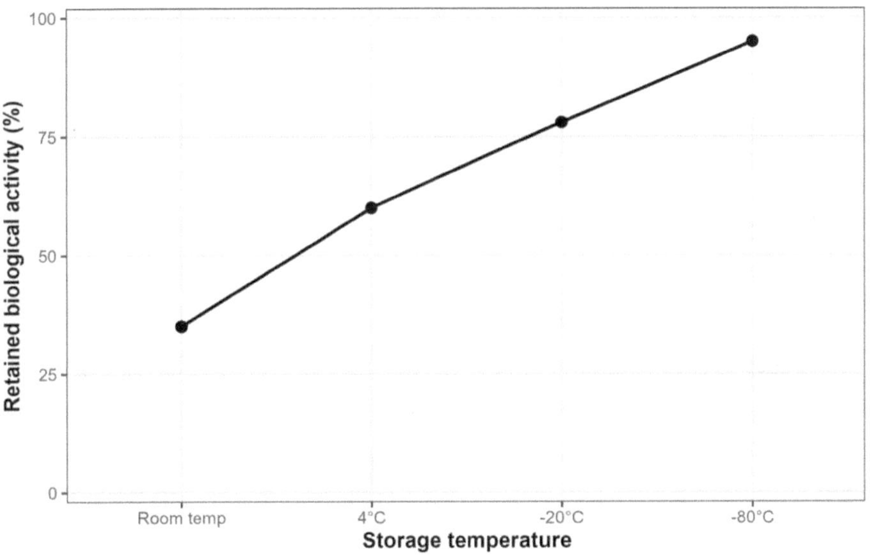

Fig. 5.6 Stability data summary chart showing the relationship between storage temperature and retained MSC–EV biological activity. Summary chart illustrating the conceptual relationship between storage temperature and retained biological activity of mesenchymal stromal/stem cell–derived extracellular vesicle (MSC–EV) preparations over time. Lower storage temperatures are generally associated with improved preservation of vesicle integrity and functional potency, whereas repeated freeze–thaw cycles and higher storage temperatures may accelerate activity loss. The figure is intended as a schematic summary of commonly reported stability trends and should be interpreted in the context of formulation, buffer composition, storage duration, and assay type

shown to maintain specific functional activities in MSC–EV applications [134, 136]. Independently, lyophilization with trehalose has been reported to preserve EV morphology and in vitro biological activities after weeks of storage in a defined EV context [135].

Lyophilization is moving from concept to translational execution. A primary off-the-shelf approach for an MSC secretome product (containing EV and protein fractions) found that trehalose-supported formulations improved EV/protein recovery after lyophilization and that shelf lyophilization protocols can further enhance recovery, supporting the feasibility of cold chain–independent products when formulation and process are optimized [154]. A more recent open-access analysis in neuroinflammatory models reports that lyophilized MSC–EVs retain immunomodulatory function and can perform comparably or better than frozen counterparts in some settings, while also emphasizing donor/source variability as a continuing challenge [154, 155].

Table 5.4 Representative stability findings to support stability-indicating specifications

Study (MSC–EV context) (references)	Storage condition(s) evaluated	Duration(s)	Primary bioactivity readout(s)	Main stability-informative conclusion
Levy et al. (2023) (BM-MSC–EVs) [133]	RT, 4 °C, −20 °C, −80 °C; lyophilized at RT	1–6 weeks (plus freeze–thaw cycles)	Endothelial gap closure; LPS-stimulated macrophage IL-6 suppression; wound healing model for cargo-loaded EVs	Bioactivity broadly retained at −20/−80 °C and after lyophilization for weeks; declines at RT/4 °C depending on assay/time; lyophilization can preserve function
Budgude et al. (2021) (MSC–EVs, cryopreservation) [136]	Cryopreservation using trehalose	Study specific; cryostorage focus	Hematopoietic stem cell expansion in vitro	Trehalose-based cryopreservation maintained a defined functional activity, supporting trehalose as a functional stabilizer
Lyu et al. (2022) (stem cell EVs, trehalose lyophilization) [135]	−80 °C vs lyophilized ± trehalose (and additional conditions)	~4 weeks (reported)	In vitro anti-inflammatory and antifibrotic markers	Trehalose-supported lyophilization preserved morphology and in vitro biological activity over weeks; −80 °C best preserved physical properties
Bosch et al. (2016) (EV stabilization principle) [134]	Trehalose during freezing/handling	Study specific	Aggregation/physical stability focus	Trehalose can prevent EV aggregation during storage stress, motivating inclusion as cryo−/lyoprotectant

5.5.2 *Recommended Storage Conditions (Evidence Linked, Not Universal)*

Given the available primary evidence and consensus emphasis on reporting and controlling storage variables, a conservative, evidence-aligned strategy for MSC–EVs is the following: (i) store isolated EVs at −80 °C for longer-term preservation when feasible; (ii) use −20 °C as an intermediate option when validated for the specific product and potency assay; (iii) minimize time at 4 °C and room temperature unless explicitly validated by stability-indicating functional assays; (iv) minimize freeze–thaw cycles through aliquoting; and (v) evaluate lyophilization with validated protectants (often trehalose/sugars) to enable shelf-stable products [5, 133, 135, 154].

5.6 Strategies to Improve Stability and Bioactivity

Strategies to improve MSC–EV stability and bioactivity fall into three interacting layers: *upstream engineering of MSC state, downstream engineering of EV composition/surface*, and *formulation/delivery engineering*. Each layer can improve therapeutic indices but also introduces additional CQAs and comparability risks, reinforcing the need for MISEV-compliant characterization and potency-linked specifications [71, 131].

5.6.1 Upstream Strategies: Controlling MSC State to Control EV Function

MSC priming (hypoxia, inflammatory cytokines, 3D culture) can alter EV release rates and EV cargo, often with the goal of enhancing immunomodulatory or regenerative function. Reviews focused on immunomodulatory applications stress that priming modifies EV miRNA/protein cargo and that inter-donor variability influences priming response, so priming must be treated as a controlled manufacturing variable [81]. Primary and review literature on MSC–EV heterogeneity indicates that cell passage and culture parameters can reduce production per cell and diminish angiogenic signaling, linking upstream control directly to downstream potency and yield [81].

5.6.2 EV Engineering and Surface Modification

Approaches to increase targeting and bioactivity include genetic engineering and chemical surface modification. Reviews of EV surface modification emphasize that these strategies aim to improve targeting and reduce off-target exposure, but they require additional characterization to confirm surface display, retained membrane integrity, and absence of new impurities [156]. A broader EV–drug development review cites examples of engineered MSC–EVs using peptide–LAMP2B strategies for targeted delivery, illustrating feasibility but also implying expanded CMC requirements [226].

5.6.3 Formulation and Delivery Engineering: Lyophilization, Excipients, and Encapsulation

Lyophilization is increasingly treated as a formulation science problem, not a simple drying step. Primary work on MSC secretome products shows that protectant choice (trehalose vs mannitol and process design) changes EV/protein recovery and retained bioactivity [154]. Reviews focused on lyophilization and EV storage highlight that excipient selection and cycle design determine aggregation, residual moisture, and reconstitution performance—parameters that should be tied to potency as a stability-indicating CQA in therapeutics [154].

Encapsulation strategies can protect EVs, increase residence time, and enable sustained release, which may increase apparent potency in vivo by improving exposure at the target site. A recent EV-focused review discusses how hydrogel encapsulation can influence MSC paracrine output and EV release, supporting combined bioprocessing + delivery design as a stability/bioactivity strategy [157].

Chapter 6
MSC–EVs Versus Other Therapeutic Platforms

Abstract This chapter provides a comparative analysis of MSC–EVs with other therapeutic platforms, including MSCs, secretome-based therapies, and engineered EV systems. Differences in mechanisms, safety profiles, dosing strategies, and clinical evidence are examined. The chapter highlights the advantages and limitations of each approach and discusses translational implications for therapeutic development. Emerging clinical data and challenges in standardization and comparability are critically reviewed.

Keywords Mesenchymal stromal/stem cells · MSC-derived extracellular vesicles · Conditioned medium · Cell-free therapy · Therapeutic platforms · Comparative efficacy · Safety · Dose metrics · Biodistribution · Priming · iPSC-derived MSCs · Potency assays · Clinical translation · Manufacturing · Regulatory science

MSCs, their secretome, and MSC-derived extracellular vesicles (MSC–EVs) are best viewed as related but distinct therapeutic platforms rather than mutually exclusive products (Fig. 6.1). Live MSCs retain environmental responsiveness, conditioned medium combines soluble and vesicular mediators, and MSC–EV products provide an acellular platform with different requirements for dose definition, biodistribution, safety, and manufacturing control. Across these platforms, the central comparative considerations are biological adaptability, definition of the active component, maturity of clinical evidence, and the principle that therapeutic risks shift rather than disappear when moving from cells to acellular products [33, 158, 159].

6.1 MSCs and MSC–EVs

Mechanistically and translationally, MSCs and MSC–EVs differ most in (a) biological autonomy, (b) biodistribution constraints, and (c) controllability of product identity.

N. M. Mussin, A. Tamadon, *Mesenchymal Stem Cells and Their Extracellular Vesicles*, https://doi.org/10.1007/978-3-032-36437-1_6

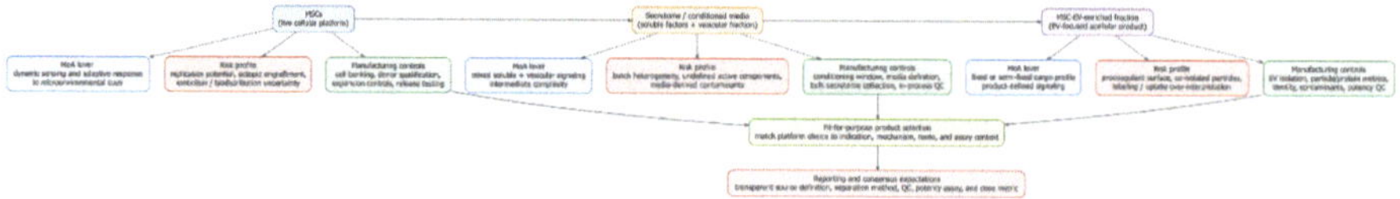

Fig. 6.1 Therapeutic platform map: MSCs to secretome/conditioned media to MSC–EV-enriched products. Therapeutic platform map illustrating the conceptual progression from live mesenchymal stromal/stem cells (MSCs) to secretome/conditioned media and finally to MSC–EV-enriched fractions as increasingly defined acellular products. The schematic compares these three therapeutic layers according to mechanism-of-action (MoA) leverage, principal risk categories, and manufacturing control requirements. Live MSCs retain dynamic environmental responsiveness but carry cell-associated risks such as replication, engraftment uncertainty, and biodistribution complexity. Secretome/conditioned media occupy an intermediate position, combining soluble mediators and vesicular components while introducing challenges related to compositional heterogeneity and undefined active fractions. MSC–EV-enriched products offer a more product-focused platform with relatively fixed cargo characteristics, but require rigorous control of isolation method, co-isolated particles, procoagulant surface features, and potency testing. The lower callouts emphasize a fit-for-purpose development strategy and transparent reporting of source, processing, characterization, functional assays, and dose metrics in line with current consensus guidance

6.1.1 Comparative Mechanisms

MSCs are living therapeutics whose functional phenotype can change in response to inflammatory cues, tissue microenvironments, and handling/cryopreservation, with therapeutic effects commonly attributed to immune modulation plus tissue-protective paracrine programs rather than durable engraftment [33].

MSC–EVs, in contrast, are a cell-free projection of MSC paracrine biology: they can transfer regulatory RNAs and proteins that influence immune and stromal cell states, but they cannot "sense and adapt" after administration in the same way cells can. This helps explain why EV effects can appear more "dose exposure dependent" with a strong dependence on route, targeting, and clearance kinetics [160].

A recurring mechanistic ambiguity is what fraction of MSC benefit is EV mediated versus mediated by soluble factors (cytokines, growth factors, metabolites) or by short-lived cell–blood interactions. High-quality inference requires fractionation and add-back designs (EV-enriched vs vesicle-depleted secretome fractions) rather than parallel single-arm experiments [160].

6.1.2 Efficacy Evidence Across Preclinical and Clinical Studies

Across organ systems, preclinical studies support the biological plausibility of MSC–EVs for inflammation resolution, barrier repair, angiogenesis, and fibrosis modulation. However, differences in EV isolation, characterization, dose metrics, and reporting continue to limit direct cross-study comparisons [5].

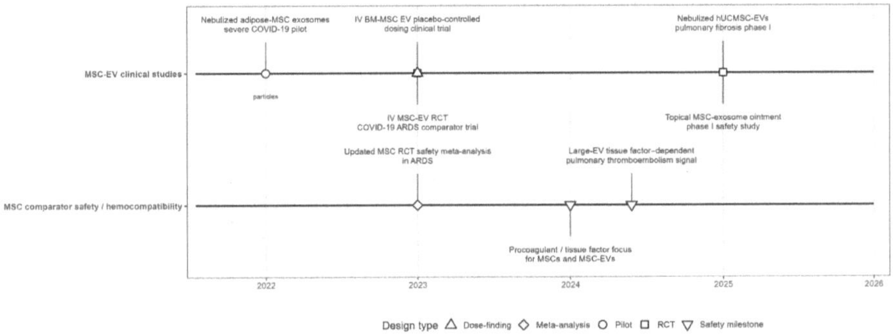

Fig. 6.2 Clinical evidence timeline for MSC–EV studies and MSC comparator safety milestones. The upper lane summarizes selected published MSC–EV studies by year, administration route, and study design. The lower lane highlights major developments in MSC safety and hemocompatibility. Symbols distinguish pilot studies, randomized trials, dose-finding studies, meta-analyses, and safety milestones

Human evidence remains less mature than the MSC cell therapy literature and consists primarily of pilot and early-phase studies, together with a small but growing number of randomized trials. Representative studies in kidney disease, respiratory disorders, pulmonary fibrosis, and chronic wounds are summarized in Fig. 6.2 and Table 6.1 [161–166]. Collectively, these reports support feasibility and short-term tolerability, but efficacy cannot yet be generalized because EV source, manufacturing process, route, dose unit, comparator, and endpoint selection differ substantially across studies [159].

6.1.3 *Safety, Dosing, Biodistribution, Immunogenicity, and Tumor Interaction*

Moving from MSCs to MSC–EVs changes rather than eliminates therapeutic risk. MSC products carry cell-associated concerns related to cell size, tissue factor–mediated coagulation, viability, culture-related variability, and potential alloimmune responses. MSC–EVs avoid cellular replication and large-cell microvascular entrapment but may retain procoagulant activity, immunomodulatory effects, and tumor interaction risks through their surface molecules and molecular cargo [167] (Table 6.1). Both platforms therefore require route-specific hemocompatibility, immunogenicity, and long-term safety assessment [42, 164, 168–170].

Dose and exposure are also defined differently. MSC products are generally administered as total cells or cells per kilogram, whereas MSC–EV studies report particle number, protein mass, or product volume, limiting dose–response comparisons. Biodistribution is strongly route dependent: intravenously administered MSCs undergo pulmonary first-pass retention, whereas systemic MSC–EVs are rapidly

Table 6.1 Clinical timeline of key EV published studies

Year	Platform	Route	Dose metric reported	Study design and signal (high level)
2016	MSC–EVs (cord blood–derived EVs used in CKD)	IV and then targeted intra-renal arterial	Protein mass (μg/kg/dose) with sequential dosing	Feasibility/safety emphasized; transient improvement signals reported; authors note a need for larger/longer studies [162]
2020	MSC–EVs (ExoFlo)	IV infusion	Product volume (mL)	Prospective nonrandomized cohort; safety and exploratory efficacy in severe COVID-19 reported [174]
2021	MSC–EVs (clinical-grade nebulized haMSC–EVs)	Nebulization	Imaging/biologic dose exploration	Preclinical biodistribution + healthy volunteer safety assessment reported as a translational step [163]
2022	MSC–EVs (haMSC–exos)	Nebulization	Particle number (nanovesicles/day)	Phase IIa single-arm inhalation trial; feasibility and tolerability emphasized; CT improvement described as hypothesis generating [171]
2023	MSC–EVs (ExoFlo)	IV infusion	Product volume (10 vs 15 mL; repeated dosing)	Multicenter randomized placebo-controlled dosing trial; safety reported; efficacy signals discussed in trial summary [161]
2024	MSC–sEVs (placental MSC–sEVs)	IV infusion	Particles/kg for 2 days	Double-blind RCT in COVID-19 ARDS reporting lower mortality in intervention arm; provides detailed particle and protein per-unit characterization and release testing [164]
2025	MSC–EVs (hUCMSC–EVs)	Nebulization	Not uniformly standardized across reports	Randomized single-blind placebo-controlled phase I in pulmonary fibrosis reporting tolerability and functional/QoL improvements with follow-up [165]
2025	MSC–EVs (WJ–MSC exosome gel)	Topical	Formulation based (gel)	Randomized double-blind controlled clinical trial in DFU reporting healing progression and reproducible preparation steps [166]

Note: This timeline reflects published clinical studies and does not equate to regulatory approval status; the heterogeneity in dose units and EV preparation remains a dominant barrier to cross-trial synthesis [159]

cleared and commonly accumulate in the liver and spleen. Inhaled, topical, intra-articular, or other local routes may increase compartment-specific exposure while reducing systemic distribution [158, 159, 163, 171–173]. Figure 6.3 summarizes these comparative exposure patterns. Detailed discussions of route-specific delivery, pharmacokinetics, dosing uncertainty, and safety assessment are provided in Chaps. 7 and 11.

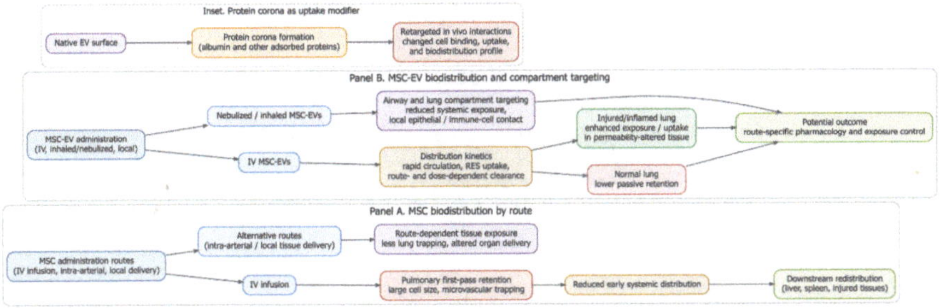

Fig. 6.3 Route-dependent biodistribution: MSCs versus MSC–EVs. A two-panel schematic illustrating route-dependent biodistribution differences between mesenchymal stromal/stem cells (MSCs) and MSC–EVs. Panel A summarizes the biodistribution behavior of MSCs after intravenous infusion, emphasizing pulmonary first-pass retention caused by cell size and microvascular sequestration, with subsequent redistribution to downstream organs and injured tissues. Alternative delivery routes, such as intra-arterial or local administration, are shown as modifiers of tissue exposure profiles. Panel B illustrates MSC–EV distribution kinetics, highlighting more rapid circulation and clearance patterns than whole-cell therapy, differential exposure in injured versus normal lung, and the compartment targeting advantages of nebulized or inhaled delivery for airway and pulmonary localization. The inset depicts protein corona formation as a mechanistic modifier of EV biodistribution, showing how adsorbed proteins such as Albumin may alter cell binding, uptake, and in vivo targeting. The figure emphasizes that route, formulation, and tissue state jointly shape exposure control and therapeutic interpretation

6.2 MSC–EVs and MSC Secretome or Conditioned Media

MSC secretome therapies sit between cell therapy and purified EV therapy. "Conditioned media" (CM) typically includes soluble cytokines/growth factors and extracellular particles; "EV-enriched secretome" aims to concentrate vesicular cargo; "vesicle-depleted secretome fraction" attempts to isolate soluble-only effects. The critical analytical issue is that these are not merely different containers of the same biology: they can differ in cargo class, stability, potency assay behavior, and scalability [160].

6.2.1 *Compositional Differences and Functional Attribution*

EVs are enriched for membrane-protected nucleic acids and proteins (and can deliver microRNAs that modulate target cell state), while vesicle-depleted fractions are dominated by soluble proteins and metabolites (Fig. 6.4). A direct fractionation study in a bleomycin lung fibrosis model illustrates why this distinction matters: in that work, intratracheal administration of MSC–EVs versus vesicle-depleted secretome fraction produced different effects depending on timing, with EVs supporting resolution of established fibrosis, while the vesicle-depleted fraction did not, and with mechanistic evidence pointing to miRNA transfer as an antifibrotic driver [160].

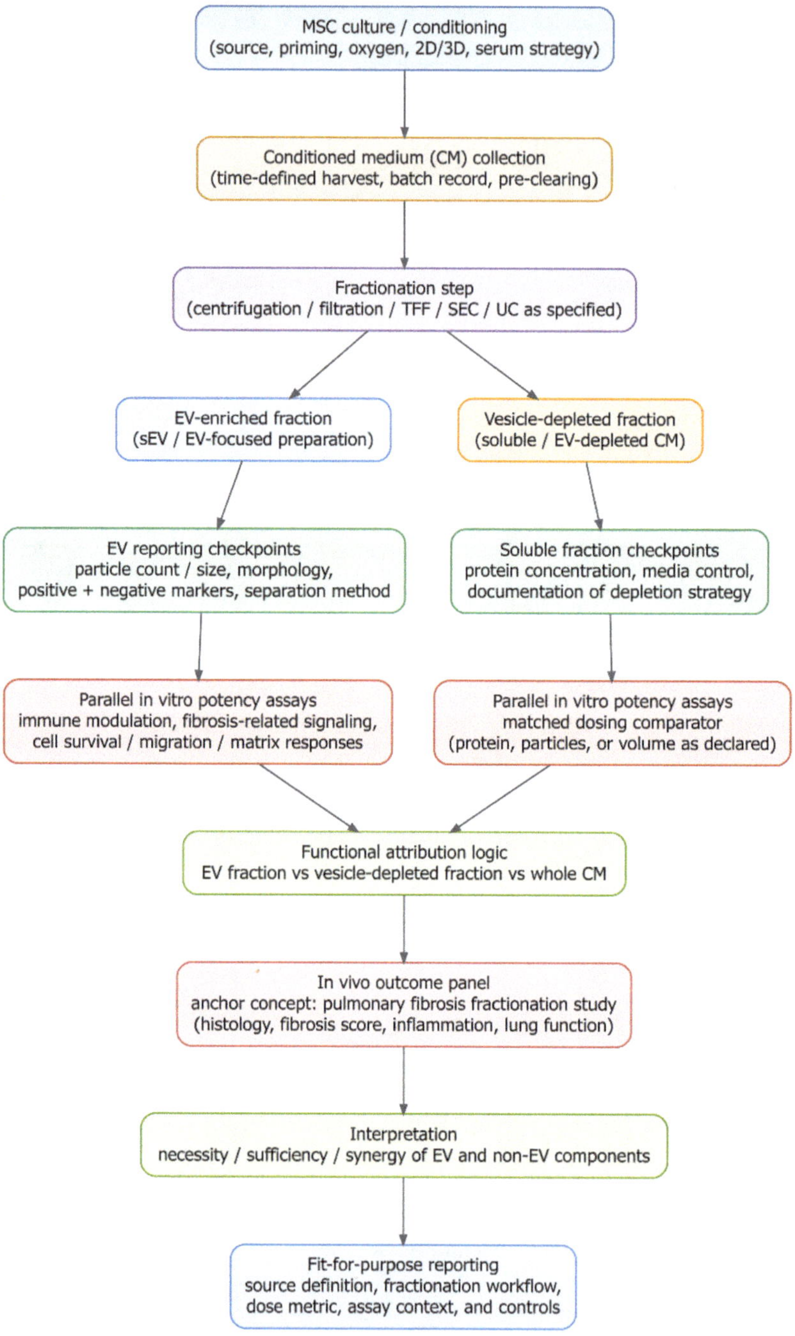

Fig. 6.4 Secretome fractionation workflow and functional attribution. Workflow diagram illustrating fractionation of Mesenchymal Stromal/Stem Cell (MSC) secretome into EV-enriched and vesicle-depleted fractions for mechanistic and functional attribution. MSC culture and conditioning are followed by conditioned-medium collection and a defined fractionation workflow using

From an experimental rigor standpoint, this kind of head-to-head fractionation is closer to the standard needed for platform comparisons than “MSC study A vs EV study B,” because it tests whether EVs are merely a purification of soluble effects or a distinct effector arm. Consensus EV guidance explicitly encourages fit-for-purpose separation and transparent reporting precisely because separation methods can co-isolate non-EV material and because functional claims require controls that rule out artifacts [5].

6.2.2 Stability, Formulation, and Storage

Secretome and conditioned-medium products may be amenable to lyophilization and relatively simple formulation, but their mixed composition makes stability assessment dependent on multiple analytes and biological activities. MSC–EVs are commonly stored frozen and require control of aggregation, freeze–thaw effects, cargo integrity, and post-thaw potency. Detailed stability evidence and recommended storage conditions are presented in Chap. 5 [164, 175].

6.2.3 Scalability and Manufacturing Trade Space

Conditioned medium may be simpler to produce because it does not require extensive particle purification, but its mixed composition complicates identity, dose definition, and potency attribution. MSC–EVs require concentration, purification, and particle characterization, yet EV enrichment may support more defined identity, dose, and functional assay frameworks. Detailed scale-up and GMP requirements are discussed in Chap. 10 [160, 164].

6.3 Primed and Unprimed MSC–EVs

“Priming” (preconditioning/licensing) aims to shift MSC–EV cargo toward a desired therapeutic profile—enhancing immunomodulation, angiogenesis, host defense, or antifibrotic signaling (Fig. 6.5). Priming is not a monolith: it is a set of

Fig. 6.4 (continued) named separation steps such as centrifugation, filtration, Tangential Flow Filtration (TFF), Size Exclusion Chromatography (SEC), or Ultracentrifugation (UC). The resulting EV-enriched fraction and vesicle-depleted fraction proceed through parallel characterization and in vitro potency assays, with explicit documentation of dose metrics and comparator logic. Downstream in vivo outcome assessment is shown as an anchor panel for fibrosis-oriented studies, illustrating how fractionation can support interpretation of EV-dependent, non-EV-dependent, or synergistic effects. Methodological callouts emphasize current extracellular vesicle reporting expectations, including transparent source definition, separation workflow, characterization, controls, and assay context

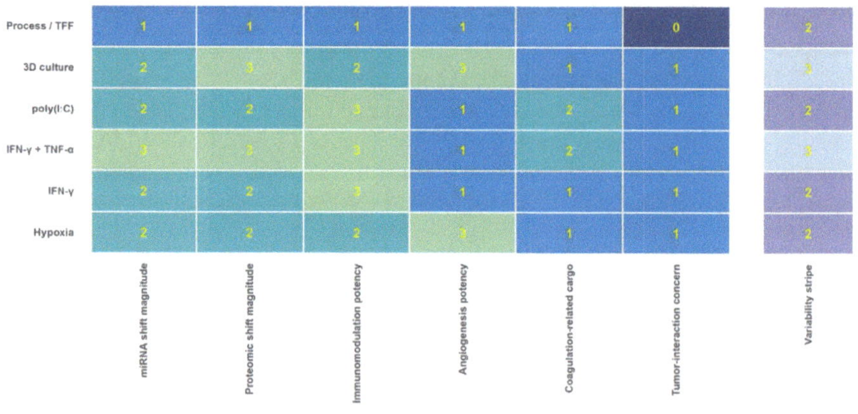

Fig. 6.5 Priming landscape heatmap: stimuli, cargo shifts, potency readouts, and safety flags. Heatmap summarizing how mesenchymal stromal/stem cell (MSC) priming or process conditions influence extracellular vesicle/secretome composition, functional potency, and selected safety-relevant signals. Rows represent priming or manufacturing-related conditions, including hypoxia, IFN-γ, IFN-γ plus TNF-α, poly(I:C), three-dimensional (3D) culture, and process-related tangential flow filtration (TFF). Columns represent semi-quantitative changes in miRNA shift magnitude, proteomic shift magnitude, immunomodulatory potency, angiogenic potency, coagulation-related cargo, and tumor interaction concern. Scores are shown on an ordinal scale from low to high effect and should be interpreted as representative summaries from selected primary studies rather than absolute cross-study comparators. The right-side variability stripe indicates the degree of donor-to-donor or batch-to-batch variation reported across conditions

interventions that can increase potency for some assays, while increasing manufacturing complexity and potentially introducing new risks.

6.3.1 Priming Modalities Used in 2016–2026 Studies

Four biological or process conditioning strategies recur in the recent literature:

Hypoxia and hypoxia mimetic priming is used to mimic ischemic/inflammatory microenvironments and can modulate EV cargo and functional outcomes. For example, EVs from hypoxic MSCs have been tested in ischemia–reperfusion contexts with reported pathway modulation in target tissues, reflecting a mechanistic rationale for hypoxia-induced cargo shifts [176].

Inflammatory cytokine priming is grounded in "licensing" MSC immune function and has been extended to EVs. A 2022 myocardial infarction study reports improved anti-apoptotic and pro-angiogenic effects of IFN-γ-primed MSC–EVs compared with unprimed EVs in vitro and in vivo, exemplifying priming as a potency amplifier [177]. In parallel, secretome-focused priming work shows that cytokine combinations can measurably enhance immunomodulatory readouts, reinforcing the broader principle that priming can shift paracrine outputs [117].

Innate immune/TLR priming can markedly reprogram EV protein cargo. A 2021 proteomic analysis found that poly(I:C) priming altered EV protein content (including enrichment of complement/coagulation cascade proteins and host defense factors), while leaving EV miRNA profiles largely unchanged, demonstrating that "priming signatures" may be protein dominant in some settings [178].

3D culture and process intensification (spheroids, 3D scaffolds, bioreactor-like systems) can increase EV yield and alter cargo. A 2025 study combining 3D MSC culture with tangential flow filtration (TFF) reported substantially higher EV extraction yield versus ultracentrifugation, while maintaining comparable in vitro activity and in vivo diabetic wound healing outcomes, illustrating that "priming" can be both biological (3D cues) and process based (isolation method) [179].

6.3.2 Molecular Changes and the "Priming Ceiling"

A key analytical finding is that priming does not uniformly rewire EV composition. A 2020 study of IFN-γ and hypoxia priming reported only limited alterations in EV miRNA landscapes with high inter-individual variability, implying that some priming methods may yield modest and donor-dependent effects for miRNA cargo—even when functional changes are observed [180]. Conversely, proteomic remodeling under TLR priming can be large, supporting a model where different priming stimuli preferentially affect different cargo layers (proteins vs miRNAs) [178].

For translation, these findings argue for a priming–potency–risk triad: priming should be selected and justified based on (i) assay-linked potency improvements relevant to the intended clinical indication, (ii) stability of the priming signature across donors/lots, and (iii) evaluation of new risks introduced by altered cargo classes (including coagulation-related proteins) [178].

6.3.3 Functional Outcomes and Emerging Risk Considerations

Priming has demonstrated functional benefits in multiple axes: cardioprotection and angiogenesis (IFN-γ-primed EVs in MI), antimicrobial/host defense factor enrichment (poly(I:C) priming), and enhanced wound healing programs when combining 3D culture with scalable purification [177]. However, priming may also amplify undesirable biological effects. Recent cancer-focused reviews indicate that MSC-derived EVs can promote tumor-associated phenotypes in a source-, state-, and cargo-dependent manner, supporting indication-specific safety assessment of primed products [167].

6.4 iMSC-Derived EVs and Adult MSC-Derived EVs

iPSC-derived MSCs (iMSCs) offer a manufacturing concept: use iPSC lines as a renewable source to produce MSC-like cells at scale and then harvest EVs as a consistent cell-free product. Adult/perinatal tissue MSCs (bone marrow, adipose, Wharton's jelly, placenta) remain the dominant sources in trials and commercial development.

6.4.1 Source Differences and Potency Implications

A 2024 study directly compared iMSCs and hUCMSCs and their EVs using a cell-based immunomodulatory assay platform (T-cell proliferation inhibition, macrophage polarization, and scratch–wound assays). It reported that iMSCs and iMSC–EVs showed immunomodulatory and pro-regenerative potential comparable to hUCMSC counterparts and that inflammatory cytokine priming enhanced the immunomodulatory potency of iMSC–EVs [181]. This is a particularly translationally relevant design because it evaluates iMSC–EVs not only in isolation but also relative to a widely used perinatal MSC source, while explicitly linking EV potency to assay readouts that could feasibly be adapted to release testing frameworks.

Beyond "small EVs," iMSC-derived large EVs have also been evaluated. A 2022 tendinopathy study reported that iMSC–lEVs mitigated pain and regulated macrophage heterogeneity in rats, with proteomic analysis identifying thousands of proteins and proposing mechanistic delivery of phosphatase regulators to macrophages with validation in patient explant tissue [182]. This suggests that iMSC-based manufacturing could support multiple EV subtypes with distinct cargo/activities but also intensifies the need to specify EV subtype and isolation method in potency and safety assessments.

6.4.2 Molecular Composition during Differentiation

Proteomic profiling provides mechanistic insight into why iMSC-derived EVs might differ from pluripotent cell EVs—and why residual pluripotent signatures matter. A 2018 differentiation study comparing iPSC EVs and PSC-derived MSC (PD–MSC)–EVs found distinct proteomic programs: iPSC EVs were enriched in proteins related to RNA/miRNA stability and protein sorting, whereas PD–MSC–EVs were richer in extracellular matrix organization, locomotion, and cell substrate adhesion pathways [183]. The translational inference is that EV identity and safety are intimately tied to differentiation state, making cell identity assays + EV cargo assays critical for iMSC–EV platforms.

6.4.3 Manufacturing Considerations for iMSC–EVs

iMSC platforms may improve upstream consistency by using a defined, renewable producer cell source, but they require additional controls for iPSC bank stability, differentiation consistency, genetic integrity, and residual pluripotent signatures. These attributes must be linked to EV identity and potency rather than assuming that a renewable producer line automatically generates a uniform EV product [181, 183].

6.5 Advantages, Disadvantages, and Translational Implications

Table 6.2 summarizes the principal comparative attributes of MSC-based therapeutic platforms.

6.5.1 Manufacturing, QC, Potency Assays, and Storage

Manufacturing control differs substantially across platforms. MSC products require donor qualification, cell banking, controlled expansion, viability, phenotype, and post-thaw functional testing. MSC–EV and iMSC–EV products require defined upstream culture conditions, scalable purification, impurity control, and mechanism-linked potency assays, whereas secretome or conditioned-medium products require simultaneous control of soluble and vesicular components.

Storage is likewise platform specific: MSCs are constrained by cell viability and post-thaw recovery, MSC–EVs by aggregation and functional drift, and secretome products by preservation of a complex mixture. Table 6.3 summarizes these comparative requirements. Detailed analytical, stability, and GMP considerations are provided in Chaps. 5 and 10 [5, 69, 164, 175, 179, 181, 184].

6.5.2 Regulatory Pathways and Clinical Translation

Regulatory classification varies according to jurisdiction and product type. MSC products may fall under cell/tissue or full biologics/advanced therapy medicinal product pathways depending on the degree of manipulation and intended use. MSC–EV and secretome products generally require biologics-like characterization, despite the absence of a fully harmonized EV-specific regulatory pathway. Across all platforms, regulatory readiness depends on traceable source materials, controlled

Table 6.2 Cross-platform comparison matrix

Attribute	MSCs	MSC–EVs (unprimed)	MSC–EVs (primed)	Secretome/conditioned media	iMSC–EVs
Primary controllable "active"	Living cell state and response	Particle cargo at release	Cargo + priming signature	Mixed soluble + particle content	Cargo with definable upstream cell line
Dose unit maturity	Higher (cells/kg)	Variable (particles, protein, volume)	Variable + priming dependent	Often volume or protein; hard attribution	Variable; can adopt particle-based dosing
Biodistribution control	Route limited; lung first-pass common IV	Route and injury state dependent	Same + possible corona/cargo effects	Route dependent; unclear dominant effector	Similar to EVs; depends on EV subtype
Acute intravascular thrombosis concern	Meaningful in some settings (TF, route effects)	Possible (PS/TF on EV surface)	Possibly increased if priming enriches coagulation proteins	Depends on composition; can include EV-mediated risk	Same as EVs; depends on preparation
Tumor interaction concern	Context dependent; culture/genomic stability considerations	Context-dependent cargo effects	Potentially increased if priming induces pro-tumor cargo	Mixed; hard to attribute	Differentiation state dependence; requires upstream controls
Storage/logistics	Cryopreserved living product; viability drift	Frozen aliquots; stability depends on handling	Same + priming signature stability	Lyophilization feasible but composition drift with storage	Same as EVs; benefits from consistent upstream source
Regulatory precedent	More developed across indications	Emerging, uneven	Emerging, uneven	Less established as a defined "drug"	Emerging; upstream complexity

Note: This qualitative comparison synthesizes evidence concerning MSC safety, EV procoagulant activity, protein corona–dependent biodistribution, clinical dose heterogeneity, secretome stability, and iMSC–EV potency [42]

TF tissue factor, *PS* phosphatidylserine, *iMSC* induced pluripotent stem cell–derived mesenchymal stromal/stem cell

Table 6.3 Minimum comparative release and characterization elements across MSC-based therapeutic platforms

Domain	MSCs	MSC–EVs/iMSC–EVs	Secretome/CM
Identity	Cell phenotype panel; viability	EV markers + negative controls; size distribution; particle/protein metrics	Defined analyte panel (cytokines/growth factors) + process controls
Purity and contaminants	Sterility, mycoplasma, endotoxin; adventitious agent risk	Sterility, mycoplasma, endotoxin; residual process reagents	Sterility, mycoplasma, endotoxin; excipient integrity
Hemocompatibility	Tissue factor/coagulation risk assessment when intravascular	Procoagulant activity assays (PS/TF related) as warranted	Depends on EV content; evaluate if used intravascularly
Potency	Indication-aligned functional assays	Indication-aligned assays (T-cell suppression, macrophage polarization, target cell repair)	Same; must separate EV vs soluble contributions when possible
Stability	Cryostability and post-thaw function	Freeze–thaw and storage impact on function; formulation effects	Storage temperature/time dependence; lyophilization excipient effects

Note: Comparative elements are based on MISEV guidance, GMP-oriented EV manufacturing frameworks, studies of procoagulant activity, and evidence concerning secretome and EV stability
CM conditioned medium, *PS* phosphatidylserine, *TF* tissue factor

manufacturing, defined identity and potency, appropriate safety testing, interpretable dose units, and well-designed clinical trials. Detailed regional regulatory requirements and lifecycle management considerations are discussed in Chap. 10 [3, 32, 185].

6.5.3 Practical Implications and Decision Logic

The central translational choice is not "MSCs or EVs," but rather which platform best matches the indication's dominant constraints (Table 6.3):

- If dynamic sensing and a sustained paracrine response are essential, MSCs may be preferred, but biodistribution, hemocompatibility, and live product logistics become primary constraints [186].

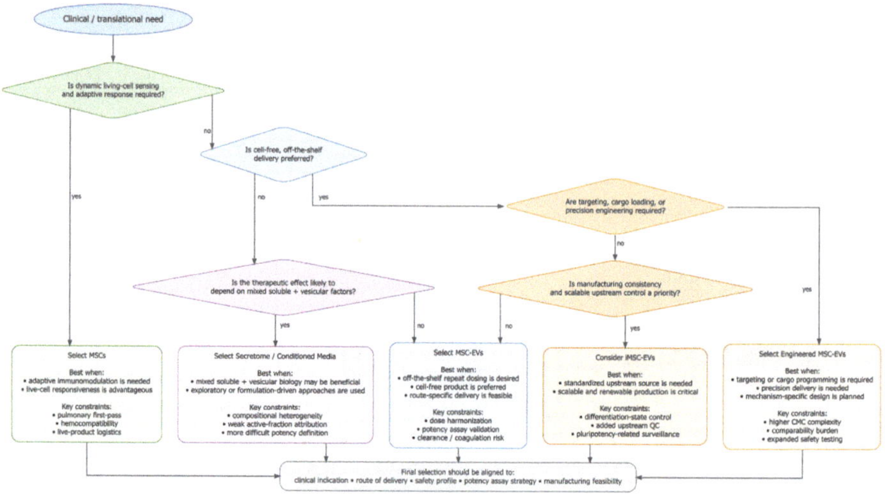

Fig. 6.6 Decision algorithm for selecting MSC-based therapeutic platforms. Flowchart guiding selection between MSCs, MSC–EVs, secretome/conditioned medium, generic MSC–EVs, and engineered MSC–EVs based on clinical indication, delivery constraints, safety profile, dynamic sensing requirements, manufacturing feasibility, and the need for targeting or cargo engineering

- If off-the-shelf availability, repeat dosing, or local administration is a primary requirement, MSC–EVs may be advantageous, but dose harmonization, potency validation, and coagulation or clearance risks must be addressed [171].
- If upstream consistency and scalability are dominant priorities, iMSC–EVs offer a renewable producer cell strategy, but differentiation state control and molecular identity surveillance remain essential [181].
- If reduced cold chain dependence is a priority, lyophilized secretome or conditioned-medium products may be attractive, but composition drift and uncertain attribution of biological activity must be controlled [175].

This comparative decision logic is summarized in Fig. 6.6.

Chapter 7
Delivery Strategies and Bioengineering of MSC–EVs

Abstract This chapter focuses on delivery strategies and bioengineering approaches for MSC-derived extracellular vesicles. It evaluates different routes of administration, including systemic and local delivery, and their impact on biodistribution and efficacy. Biomaterial-assisted delivery systems such as hydrogels and scaffolds are discussed as strategies to enhance retention and targeting. The chapter also reviews engineering approaches for cargo loading and surface modification to improve therapeutic performance.

Keywords EV delivery · Biomaterials · Targeting · Bioengineering · Drug delivery systems

MSC–EVs are being developed as "cell-free" biologics whose therapeutic activity depends not only on cargo composition but also on how, where, and for how long EVs are delivered in vivo. Across the 2016–2026 literature, the dominant delivery challenge is not whether MSC–EVs can elicit anti-inflammatory, pro-angiogenic, and cytoprotective effects, but rather how to achieve sufficient spatiotemporal exposure at the target tissue while managing off-target biodistribution, rapid clearance, and batch-to-batch variability [5].

Evidence from comparative biodistribution studies shows that systemic IV delivery concentrates MSC–EV signals in the liver and spleen (and often lungs), whereas local delivery routes can confine EV signal to the intended organ with reduced peripheral exposure [187]. Yet, efficacy depends on disease context: in injured lungs, MSC–EV uptake can increase and persist, indicating that pathological microenvironments may create "demand-driven" retention.

Biomaterial-assisted delivery (hydrogels, electrospun scaffolds, adhesive depots, and microneedle-based systems) repeatedly demonstrates a mechanistic advantage: protect EVs, slow their loss, and reshape local immune responses. Quantitatively, an exosome-functionalized electrospun scaffold retained ~50.8% of local exosome fluorescence at day 7 (whereas injected EVs were undetectable after 24 h) and improved large wound closure to ~65.1% at 2 weeks (vs ~29.2% for injected EVs) [188]. In myocardium, alginate hydrogel embedding of MSC–sEVs supported ~10-day in vitro release and improved cardiac outcomes versus free sEV injection after

N. M. Mussin, A. Tamadon, *Mesenchymal Stem Cells and Their Extracellular Vesicles*, https://doi.org/10.1007/978-3-032-36437-1_7

MI [189]. In spinal cord injury, a porous microneedle patch enabling sustained local EV delivery for at least 7 days yielded meaningful functional gains: 4/11 rats achieved BBB locomotion scores >10 at 8 weeks, a distribution not observed in comparators [190].

Bioengineering (cargo loading and EV surface modification) is increasingly used to overcome delivery bottlenecks. Endogenous cargo enhancement via producer cell engineering can be highly effective: miR-17-5p-engineered MSC-sEVs increased miR-17-5p expression ~16-folds, and GelMA encapsulation achieved ~97.24 ± 3.47% wound closure by day 15 in diabetic mice while maintaining biocompatibility by histology [191]. Targeting ligands can improve deep tissue access: CAP-displaying EVs (Lamp2b fusion) fused with liposomes delivered CRISPR plasmids into chondrocytes and reduced OA severity with OARSI scores ~1.11 vs ~12.89 in untreated DMM rats [192].

For translation, the field's limiting factors are increasingly standardization, potency correlation, scalable manufacturing, and route-specific safety—themes emphasized by the MISEV2018 and MISEV2023 consensus guidelines [5]. Clinical translation is active but heterogeneous, with trials exploring IV infusion, intra-articular injection, and nebulized inhalation registered in ClinicalTrials.gov [188], underscoring the need for route-specific dosing logic and validated release/retention claims [193]. This chapter synthesizes primary research articles and high-quality reviews, emphasizing mechanistic delivery biology, quantitative in vivo evidence where reported, and translational implications.

"MSC–EVs" is used broadly to include small EVs/exosome-enriched preparations derived from MSCs (bone marrow, adipose, umbilical cord, etc.). Nomenclature varies across papers; therefore, interpretation is anchored to each study's isolation/characterization approach and is assessed through the lens of MISEV guidelines [5].

Quantitative "dose" is reported heterogeneously: particle number (NTA), protein mass (μg), or product volume. When studies report only one metric, cross-study comparisons are treated as semi-quantitative, and limitations are made explicit [71]. Finally, EV tracking (fluorescent lipophilic dyes, genetic reporters, etc.) can misrepresent true EV fate due to dye transfer, label persistence, and altered EV properties. When biodistribution is a central claim, this chapter prioritizes studies that explicitly acknowledge these constraints [187].

7.1 Routes of Administration and Dosing Paradigms

MSC–EV administration is best framed as route-dependent pharmacology (Fig. 7.1): the same EV product can yield distinct tissue exposure profiles and mechanisms of action depending on how it enters the body. A 2023 comparative biodistribution study administering DiR-labeled MSC-EVs to mice at 2.00 × 10^10 EV per mouse directly demonstrated route-governed localization: IV delivery concentrated signal in the abdominal region and, ex vivo, in the spleen and liver; intratracheal delivery

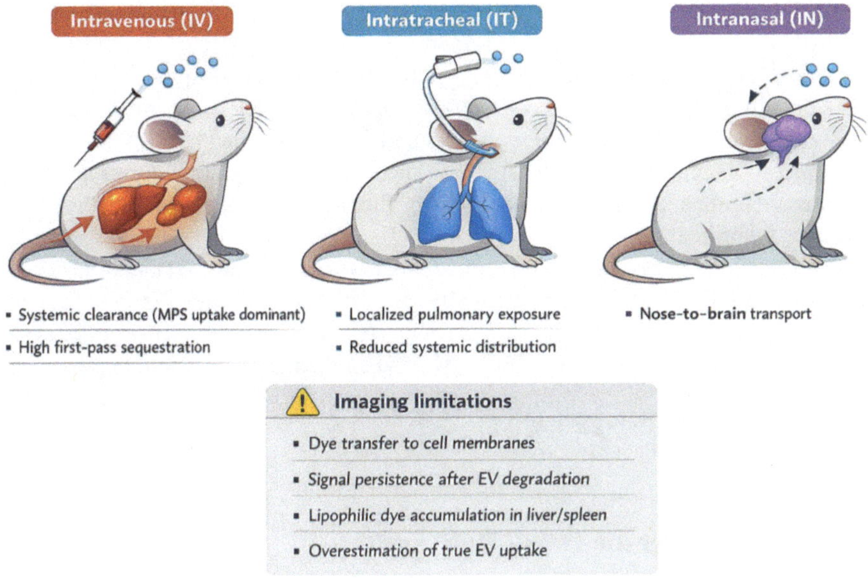

Fig. 7.1 Route-dependent biodistribution of MSC-derived extracellular vesicles (MSC–EVs): conceptual organ "hot spots" and exposure constraints. Schematic illustration of how the route of administration governs the in vivo biodistribution of mesenchymal stromal/stem cell–derived extracellular vesicles (MSC–EVs). Intravenous (IV) delivery results in predominant accumulation within the mononuclear phagocyte system, particularly in the liver and spleen, with secondary distribution to the lungs. Intratracheal (IT) administration achieves localized pulmonary exposure with minimal systemic dissemination. Intranasal (IN) delivery enables preferential transport to the brain via olfactory and trigeminal pathways, supporting nose-to-brain targeting. The figure also highlights major methodological limitations in EV tracking studies, including dye transfer to cell membranes, persistence of signal after EV degradation, preferential accumulation of lipophilic dyes in reticuloendothelial organs, and potential overestimation of true EV uptake. This conceptual model integrates findings from comparative preclinical studies and emphasizes the importance of route selection in optimizing therapeutic exposure and minimizing off-target distribution

concentrated signal in the lungs; intranasal delivery concentrated signal in the brain after 24 h [187].

Beyond the route itself, measurement strategy matters. The same comparative study explicitly notes limitations of lipophilic dyes (membrane transfer, long half-life, potential signal persistence after EV degradation, and altered biodistribution toward lipid-rich organs), meaning "signal localization" should not be equated with intact EV uptake without orthogonal validation [187].

Dose levels in preclinical MSC–EV studies commonly land in the realm of tens of micrograms protein or ≥10^9 to 10^10 particles per mouse/rat per administration, but with wide variability. Examples include the following: (i) 2.00 × 10^10 EV/mouse for biodistribution mapping across IV/IT/IN routes [187], (ii) 2 × 10^9 particles per dose for either intramyocardial or IV delivery in a rat MI study [187], and (iii) 80 µg sEV protein mass delivered intramyocardially (with or without alginate hydrogel) at 30 min post-MI in rats [189].

In local/tissue resident delivery (skin wounds, joint space, myocardium, spinal cord), dosing is often "depot based": EVs are applied, immobilized, or gelled in situ to maximize exposure time rather than increasing peak concentration. A scaffold immobilization approach used 30 µg exosome protein in standard wounds and 50 µg for large wounds and separately used 15 µg fluorescently labeled EVs for in vivo retention quantification [188].

Mechanistically, these dose selections reflect two constraints: (i) clearance dominates for free EVs (especially systemically) and (ii) "effective dose" is a function of AUC at target tissue, not simply injected quantity. The translational corollary is that route-specific dosing should be designed around measurable exposure surrogates (retention imaging, tissue EV uptake, pathway activation) rather than around protein mass alone [71] (Table 7.1 and Fig. 7.2).

7.2 Systemic Versus Local Delivery

Systemic delivery (most commonly IV) is attractive clinically because it is scalable and repeatable and does not require invasive access to target tissues (Fig. 7.3). However, systemic delivery faces a fundamental biological filter: rapid recognition and uptake by the mononuclear phagocyte system, often yielding dominant signals in the liver and spleen. The IV arm of the 2023 IV/IT/IN comparison showed ex vivo enrichment in the spleen and liver at 24 h [187]. Similarly, in an acute lung injury (ALI) model, enriched lung uptake depended strongly on injury context and timing rather than purely on dose, illustrating that "targeting by pathology" can partially compensate for systemic dilution [194].

Local delivery routes can overcome the systemic filter by confining initial exposure to the desired site. Intratracheal delivery of MSC–EVs, for example, can yield selective lung accumulation with minimal peripheral spread at 24 h; intranasal delivery in the same study produced a signal largely confined to the brain [187]. Conceptually, these results support a compartment-first model: deliver EVs into the anatomical compartment closest to the intended cellular targets whenever feasible, and then rely on diffusion/transport locally rather than on systemic homing [187].

Efficacy comparisons between systemic and local delivery are disease and model dependent. A 2023 rat MI study delivering EVs either intramyocardially or intravenously (2 × 10^9 particles) reported no significant improvement in cardiac function or infarct size and observed no significant group differences in derived measures of LV function at 28 days post-MI, underscoring that "local delivery" alone does not guarantee efficacy if the EV product, timing, or dose is insufficient (or if tracking fails to confirm effective myocardial exposure) [187]. Conversely, in spinal cord injury, a local, sustained-delivery microneedle patch approach achieved substantial functional improvements over controls, implying that in highly compartmentalized

Table 7.1 Comparative overview of MSC–EV delivery routes: biodistribution, advantages, limitations, and dosing paradigms

Route	Target compartment	Practical advantages	Key limitations	Biodistribution “signature”	Typical preclinical doses (examples)	Notes on monitoring/ labeling pitfalls
Intravenous (IV)	Systemic circulation → liver, spleen (MPS), lung (secondary)	Clinically feasible; minimally invasive; scalable; repeat dosing possible	Rapid clearance by mononuclear phagocyte system (MPS); low target specificity; high off-target exposure	Strong enrichment in the liver and spleen; moderate lung signal; minimal brain penetration	~2.00×10^{10} EV/ mouse (comparative biodistribution study); ~2×10^{9} particles (rat MI models)	Lipophilic dyes accumulate in the liver/spleen → overestimation of uptake; signal persistence after EV degradation; difficult-to-distinguish intact EVs vs debris
Intratracheal (IT)	Lung (airway and alveolar compartment)	Direct delivery to the lung; high local concentration; reduced systemic exposure; suitable for pulmonary diseases	Requires airway access; limited to respiratory indications; uneven distribution in diseased lung	Highly localized lung enrichment; minimal systemic spread	~2.00×10^{10} EV/ mouse (comparative study); lower doses often effective due to compartment targeting	Fluorescent signal may reflect surface adherence rather than cellular uptake; inflammation-dependent uptake may bias interpretation
Intranasal (IN)	Brain via olfactory/ trigeminal pathways	Non-invasive; enables CNS targeting; bypasses blood–brain barrier (BBB)	Variable efficiency; species-dependent translation; limited dose volume; uncertain deep brain penetration	Preferential brain localization; low peripheral organ signal (in rodents)	~10^{9} to 10^{10} EV/ dose (typical experimental range; comparative biodistribution studies)	Overinterpretation of “nose-to-brain” delivery; dye diffusion along mucosa; inconsistent tracking across labeling methods

(continued)

Table 7.1 (continued)

Route	Target compartment	Practical advantages	Key limitations	Biodistribution "signature"	Typical preclinical doses (examples)	Notes on monitoring/ labeling pitfalls
Intramyocardial (local injection)	Myocardium (injured cardiac tissue)	High local delivery; bypasses systemic clearance; precise targeting	Invasive; rapid washout due to cardiac motion; limited retention without biomaterials	Local cardiac signal initially, but rapid decline without retention strategies	~2 × 10^9 particles or ~80 μg EV protein (rat MI studies)	Signal loss may reflect washout rather than degradation; injection leakage may confound localization
Topical/scaffold based (biomaterial assisted)	Local tissue	Sustained release; enhanced retention; protection of EVs; immune modulation	Requires biomaterial optimization; limited systemic reach; regulatory complexity	Prolonged local retention; minimal systemic distribution	~15 to 50 μg EV protein (wound/ scaffold studies)	Fluorescence persistence reflects depot retention rather than active uptake; scaffold autofluorescence must be controlled

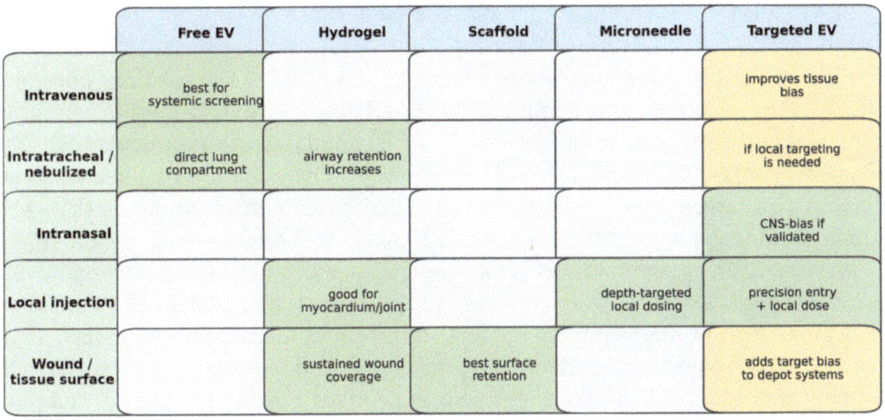

Fig. 7.2 Route–platform fit map for MSC–EV delivery strategies. Matrix comparing how different delivery platforms align with major administration routes including intravenous, intratracheal/nebulized, intranasal, local injection, and wound/tissue surface application. Green zones indicate the strongest current conceptual fit, yellow zones indicate conditional or context-dependent use, and white zones indicate weaker or less established combinations. The framework highlights that optimal EV delivery depends on matching platform design to anatomical compartment, retention needs, and translational feasibility rather than treating all routes as interchangeable

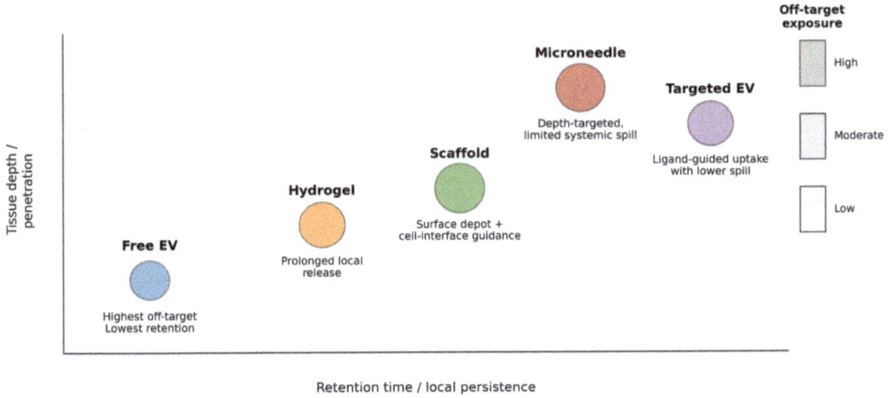

Fig. 7.3 Exposure engineering profiles across MSC–EV delivery platforms. Comparative schematic of free EVs, hydrogel-based delivery, scaffold-based delivery, microneedle-assisted delivery, and targeted EVs plotted across three translational dimensions: retention time, tissue depth/penetration, and off-target exposure. Free EVs show rapid dispersion and the highest systemic spill, whereas biomaterial-assisted platforms increase local persistence. Microneedle systems improve depth-targeted deposition, and targeted EVs enhance uptake bias while potentially reducing off-target exposure. The figure emphasizes that delivery strategy changes therapeutic exposure by reshaping where EVs go, how long they remain, and how selectively they enter target tissues

tissues with hostile microenvironments, sustained local exposure may be more decisive than bolus route selection [190].

Biodistribution is not only an exposure question; it is also a safety question. Local delivery can reduce systemic exposure and thus may reduce risks of off-target immunomodulation, ectopic effects, or organ accumulation. Intra-articular delivery of engineered CAP-decorated hybrid EVs showed signals confined to the knee joint, whereas non-targeted hybrid EVs showed signal in the kidney, implying that even within a nominally local route, surface targeting can reshape off-target escape [192].

From a dosing standpoint, systemic routes often require higher doses to overcome first-pass clearance, whereas local routes can use lower total doses but demand strategies to prevent washout (joint fluid turnover, cardiac motion/pressure, CSF flow). The convergence of evidence supports a practical translational heuristic: treat systemic dosing as a distribution problem (how much reaches the target), and treat local dosing as a retention problem (how long it stays) [187].

7.3 Biomaterial-Assisted Delivery

Biomaterial-assisted delivery aims to transform MSC–EV administration from a transient bolus into a controlled, tissue-interactive therapy. Three mechanistic functions recur across successful designs (Fig. 7.4): (i) retention (physical entrapment/immobilization reduces convective loss), (ii) protection (stabilization against enzymatic and oxidative attack), and (iii) instruction (the material shapes immune cell recruitment and phenotype, amplifying EV signaling) [188].

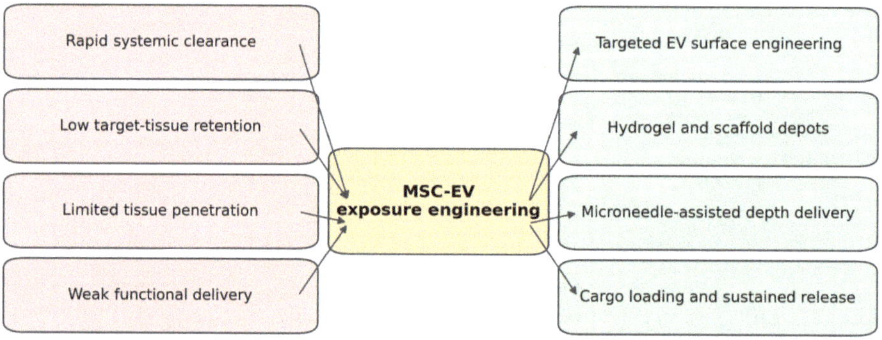

Fig. 7.4 Delivery barriers and corresponding engineering solutions for MSC–EVs. Conceptual framework linking major delivery barriers—rapid systemic clearance, low target tissue retention, limited tissue penetration, and weak functional delivery—to the principal engineering approaches developed to address them. Targeted EV surface engineering is positioned against non-specific clearance, hydrogels and scaffolds against local washout, microneedles against shallow tissue access, and cargo loading plus sustained-release strategies against insufficient intracellular effect. The figure emphasizes that dose escalation alone does not solve the exposure problem; instead, different barriers require different engineering layers

A scaffold immobilization exemplar is the Science Advances 2021 exosome-functionalized electrospun scaffold study [187]. In vivo, subcutaneously injected EVs became undetectable after 24 h, whereas EVs tethered to the scaffold persisted with 50.8 ± 5.8% fluorescence remaining at day 7 (relative to day 0) [188]. Cellular analysis showed that macrophages constituted ~70.4 ± 4.8% of exosome-positive cells in the wound tissue, indicating that the scaffold acted as both a depot and an immune cell training platform rather than a passive reservoir [188].

The same scaffold system translated retention into tissue outcomes: in a large wound model, exosome-functionalized scaffolds achieved 65.1 ± 5.3% wound closure at 2 weeks, compared with 29.2 ± 10.5% for injected EVs and 18.2 ± 4.7% for untreated wounds, using an exosome dose of 50 μg protein mass [188]. These numbers highlight a core translational mechanism: biomaterials can convert EV signals into durable, structured immune programs (M2/Treg skewing reported in the study) that persist long enough to influence remodeling [188].

In myocardium, an alginate hydrogel approach provides complementary evidence. In vitro, 100 μL hydrogel containing 80 μg sEVs showed release over ~10 days with early burst behavior for selected cross-linking conditions [189]. The hydrogel chosen exhibited a storage modulus G′ in the ~400 to 1800 Pa range (a cardiac tissue engineering–relevant window as discussed by the authors), which reflects an explicit "mechanical matching" principle to balance injectability with retention under dynamic cardiac motion [189].

In vivo, the same study delivered 80 μg sEVs intramyocardially with or without hydrogel at 30 min post-MI and observed stronger heart retention signals at days 3 and 7 with hydrogel versus free sEVs, alongside reduced peripheral signals (liver/spleen) at day 3—consistent with a retention-dominant mechanism [189]. While many outcome details are figure based, the study reports improved LVEF/LVFS and reduced infarct size with sEV hydrogel versus free sEV treatment at 28 days, indicating functional translation of sustained local exposure [189].

7.4 Hydrogels, Scaffolds, and Targeted Delivery Systems

Hydrogels and scaffolds are not interchangeable; they represent different control "levers" over EV delivery. Hydrogels (especially injectable or photo-cross-linkable systems) optimize conformality and intimate tissue contact, whereas scaffolds (fibrous meshes, porous matrices) optimize surface control, cell infiltration, and long-lived depots. Targeted systems overlay molecular recognition (ligands, peptides) onto either platform [189].

Mussin et al. [195] highlighted exosome-infused hydrogels as a strategy to protect vesicles, increase local retention, and provide sustained release at injured tissues. Translation of these systems requires simultaneous control of EV identity and potency, loading efficiency, spatial distribution within the hydrogel, release kinetics, material degradation, sterilization, storage stability, and the comparability of the combined EV biomaterial product.

Design principles emerge clearly from the highest information studies:

Mechanical and structural matching is essential. In cardiac delivery, hydrogel stiffness was explicitly tuned by $CaCl_2$ concentration to avoid being "too soft" (poor retention) or "too hard" (delivery difficulty/mismatch), aligning G′ to a cardiac-relevant window [189]. In skin and wound dressings, GelMA systems are selected for porous structure, swelling, and barrier properties; local hydration and oxygen/nutrient exchange are framed as supportive mechanisms for cellular migration and proliferation [196].

Release kinetics must be engineered, not assumed. The alginate sEV hydrogel showed an ~10-day release with burst early days for selected formulations [189]. In an engineered GelMA system delivering miR-17-5p-enriched sEVs, free sEV fluorescence in recipient cells decayed by 72 h, whereas GelMA-mediated delivery maintained stable signal at 72 h in vitro; in vivo, injected free sEVs were "metabolized almost completely" by day 5, while GelMA-mediated fluorescence remained detectable at day 5 [191].

Sustained delivery can enable "therapeutic coverage" during the critical inflammatory window. In spinal cord injury, the porous microneedle patch strategy was explicitly designed around the first week post-injury as the optimal window for suppressing acute neuroinflammation. The system achieved sustained delivery for at least 7 days, and, at 8 weeks, 4/11 treated rats reached BBB >10 (sustained weight support with plantar stepping), while control groups remained largely <4 [190]. This pattern suggests that delivery systems can shift MSC–EV therapy from transient immunosuppression toward structural preservation (axon sparing and reduced cavity/scar formation described by authors) [190].

Targeted delivery systems can solve "deep tissue access" limits. Cartilage is an archetypal delivery barrier: dense matrix and low vascularity impede nanoparticle penetration. A targeted hybrid system genetically displayed a chondrocyte affinity peptide (CAP) via Lamp2b on exosome surface, fused EVs with liposomes, and used intra-articular administration to deliver CRISPR/Cas9 plasmids. Targeting enabled penetration into deeper cartilage regions and local retention for at least 7 days, while also confining signals to the joint (as opposed to kidney signal observed with non-targeted hybrid EVs) [192].

In vivo outcome linkage is strongest when biomaterial performance metrics are paired with biological readouts. In the scaffold study, retention (50.8% at day 7) paired with macrophage-dominant uptake and wound closure/angiogenesis outcomes, creating a coherent chain from depot physics → immune trafficking → tissue repair [188]. In the engineered GelMA system, EV retention and delayed clearance paired with wound closure (97.24 ± 3.47% at day 15), angiogenesis/collagen deposition markers, and absence of organ abnormalities by histology, providing a similarly coherent translational narrative (Table 7.2) [191].

Table 7.2 Biomaterial-assisted MSC–EV delivery systems: composition, release behavior, and therapeutic outcomes

Platform type	Material(s)	EV incorporation	Release profile (in vitro/in vivo)	Model	Quantitative outcomes	Key limitations
Electrospun scaffold (functionalized)	Electrospun polymer scaffold + PEI (polyethyleneimine) surface modification	Immobilization via electrostatic binding (surface tethering)	Sustained retention; ~50.8% signal retained at day 7 vs undetectable at 24 h for injected EVs	Full-thickness skin wound (rodent)	~65.1% wound closure at 2 weeks vs ~29.2% (injected EVs) and ~18.2% (control); ~70% of EV+ cells were macrophages	Limited control over release kinetics; potential material-related cytotoxicity (PEI); surface-bound EVs may not fully internalize
Injectable hydrogel (cardiac)	Alginate hydrogel (Ca^{2+} cross-linked)	Encapsulation within hydrogel matrix	~10-day release in vitro; enhanced retention in myocardium at days 3–7 vs free EVs	Myocardial infarction (rat)	Dose: ~80 µg EVs; improved cardiac function (↑LVEF/LVFS), reduced infarct size vs free EVs	Initial burst release; mechanical mismatch may affect delivery; invasive administration required
Porous microneedle patch	Biodegradable polymer microneedles (porous structure)	Sustained diffusion from embedded EV reservoir	Continuous delivery ≥7 days in vivo (controlled release into tissue)	Spinal cord injury (rat)	Functional recovery: Subset achieved BBB score >10 at 8 weeks (vs minimal recovery in controls); reduced inflammation and scar formation	Complex fabrication; limited penetration depth; dosing uniformity challenges
Photo-cross-linked hydrogel (engineered EV depot)	GelMA (gelatin methacryloyl) hydrogel + engineered MSC–sEVs	Encapsulation (3D hydrogel network)	Sustained release; detectable EV signal at day 5 vs near-complete loss of free EVs; stable in vitro uptake up to 72 h	Diabetic wound healing (mouse)	~97.24% wound closure by day 15; enhanced angiogenesis and collagen deposition; no major organ toxicity	UV/photo-cross-linking may affect EV integrity; hydrogel degradation variability; scale-up challenges

(continued)

Table 7.2 (continued)

Platform type	Material(s)	EV incorporation	Release profile (in vitro/in vivo)	Model	Quantitative outcomes	Key limitations
Topical hydrogel/ scaffold hybrid systems (generalized)	Natural/synthetic hydrogels (GelMA, collagen, chitosan) ± scaffold support	Encapsulation + partial adsorption	Tunable release (days to weeks depending on cross-linking and porosity)	Skin, bone, cartilage regeneration models	Improved tissue regeneration, angiogenesis, and reduced inflammation compared to free EVs	Batch variability; difficulty standardizing release kinetics; regulatory complexity

7.5 Cargo Loading and Engineered MSC–EVs

Engineering MSC–EVs can be decomposed into two orthogonal goals (Fig. 7.5): (i) increase the amount or functionality of therapeutic cargo and (ii) increase delivery efficiency to the right cells and subcellular compartments. Engineering methods are broadly categorized as endogenous (producer cell engineering) and exogenous (post-isolation loading), each with different efficiency and safety constraints [197].

Endogenous engineering is increasingly favored for nucleic acid cargo enrichment because it leverages natural EV biogenesis. A high-information example is miR-17-5p overexpression in umbilical cord MSCs via lentiviral modification, producing engineered sEVs with an ~16-fold increase in miR-17-5p and preserved EV morphology/markers [191]. When these engineered sEVs were loaded into photo-cross-linked GelMA and applied to diabetic wounds, the system achieved 97.24 ± 3.47% wound closure by day 15, coupled with mechanistic targeting of PTEN and p21 pathways (senescence and proliferation control), and favorable safety readouts (no organ histologic abnormalities or immune infiltration in major organs) [191].

Exogenous loading can be operationally simpler but often suffers from low encapsulation efficiency and potential EV damage or aggregation. A 2024 high-quality review reports that electroporation can be extremely low efficiency (<0.05% encapsulation efficiency) in some contexts and highlights the need to validate that the cargo is truly intraluminal rather than adsorbed/aggregated [197]. Despite efficiency concerns, electroporation is still used in translational designs: a 2023 microneedle patch strategy loaded human umbilical cord MSC–EVs with miR-29b mimics via electroporation to target cardiac fibroblasts and suppress fibrosis-associated proteins (reported at abstract level) [198].

Hybrid systems (exosome–liposome fusion, surface-displayed ligands, and cargo packaging aids) are emerging to address both loading and targeting. In osteoarthritis, hybrid CAP–EVs were used to encapsulate CRISPR/Cas9 plasmids and achieved strong in vivo outcomes: OARSI scores approximately 1.11 in the targeted CRISPR group versus 12.89 in untreated DMM OA rats, indicating near-complete recovery by histological scoring [192]. Notably, toxicity readouts showed the liposome formulation had ~65% viability in chondrocytes and ~3% hemolysis, whereas EVs and

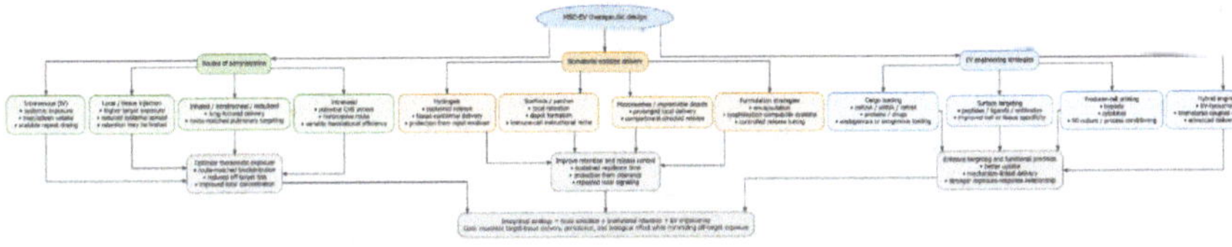

Fig. 7.5 Integrated delivery and engineering landscape for MSC–EVs. Schematic combining routes of administration, biomaterial-assisted delivery, and EV engineering strategies to optimize therapeutic exposure and targeting

hybrid vesicles showed minimal cytotoxicity and hemolysis—evidence that hybridization can "buffer" liposome toxicity while retaining loading advantages [192].

Safety must be treated as a multi-scale constraint: molecular (unintended cargo), cellular (off-target editing or phenotypic shifts), and systemic (immunogenicity, coagulation, organ accumulation). The OA CRISPR study explicitly foregrounds risks such as off-target effects and immune reactions in CRISPR translation and motivates targeted delivery as risk minimization [192]. In contrast, engineered sEV hydrogel systems for wound healing emphasize tissue biocompatibility and histology as the primary safety endpoint [191]. For intranasal neurodelivery, a 2026 open-access micro-report in *Molecular Brain* found that repeated intranasal MSC–sEV dosing (3 consecutive days/week from symptom onset) did not significantly extend survival but did delay progression from mild to more severe symptoms in the SOD1(G93A) ALS mouse model, reframing safety/benefit as preservation of function rather than survival extension [199].

A translationally important (and often underaddressed) safety mechanism is route-dependent off-target effect. For example, intranasal MSC–EV delivery can concentrate signals in the brain with minimal peripheral distribution in some mouse imaging studies, but primate data summarized in high-quality reviews show that intranasal delivery may yield limited plasma signal and variable CNS signal depending on reporters and dose; therefore, claims of "nose-to-brain superiority" require orthogonal validation [187] (Table 7.3 and Fig. 7.6).

7.6 Clinical Translation Barriers for Engineered MSC–EVs

Despite rapid progress in EV engineering, clinical translation of engineered MSC–EVs remains constrained by several unresolved barriers. First, engineered EVs introduce additional complexity beyond native EV products, including altered surface ligands, loaded nucleic acids or proteins, hybrid membrane structures, and formulation-dependent behavior. Each modification may improve targeting or potency but also creates new critical quality attributes that must be measured and controlled.

Second, targeting does not necessarily equal functional delivery. Increased binding to a tissue or cell type does not guarantee productive internalization, endosomal escape, or biologically meaningful cargo activity. Therefore, engineered MSC–EV products require potency assays that measure functional delivery rather than uptake alone. Third, engineering strategies may introduce safety concerns, including off-target cargo transfer, immunogenicity, residual reagents, altered biodistribution, procoagulant activity, and long-term effects of gene regulatory cargo.

Fourth, comparability becomes more difficult as engineering complexity increases. Changes in producer cell engineering, loading method, purification platform, or formulation can alter EV identity and activity. For clinical translation, engineered MSC–EVs should therefore be developed under a lifecycle framework that integrates product characterization, potency assays, biodistribution studies, safety

Table 7.3 Cargo loading strategies for engineered MSC–EVs: methods, efficiency, functional outcomes, and safety considerations

Loading strategy	Method (endogenous/exogenous/hybrid)	Cargo type	Loading efficiency (reported range)	Functional readouts	Safety liabilities	Representative studies
Producer cell genetic engineering for miRNA enrichment	Endogenous	miRNA (example: miR-17-5p)	High relative enrichment in cargo; example: ~16-fold increase in miR-17-5p in engineered sEVs versus controls	Enhanced wound repair, angiogenesis, and collagen remodeling; in a diabetic wound model using GelMA as depot, wound closure reached 97.24 ± 3.47% by day 15	Risks from parental cell modification, unintended cargo shifts, variability in transgene expression, and manufacturing complexity	Engineered UC–MSC–sEVs enriched with miR-17-5p for diabetic wound healing
Direct electroporation loading	Exogenous	miRNAs, siRNAs, small nucleic acids, occasionally other charged biomolecules	Often variable and sometimes very low; reported to be <0.05% in some contexts	Useful when rapid post-isolation loading is desired; can support proof-of-concept delivery, but biological effect depends heavily on true intravesicular loading rather than surface adsorption or aggregation	EV membrane damage, cargo aggregation/precipitation, poor reproducibility, uncertain true encapsulation, altered vesicle size/dispersity	High-quality recent EV drug delivery reviews summarizing electroporation constraints
Electroporated EVs with therapeutic miRNA	Exogenous	miRNA mimic (example: miR-29b)	Not consistently quantified in all reports; efficiency depends on pulse conditions and purification workflow	Antifibrotic or regenerative signaling after local sustained delivery; used in advanced depot systems such as biomaterial-assisted patches	Same electroporation risks as above, plus uncertainty about retained biological potency after loading	Example summarized in recent literature: MSC–exosome loading with miR-29b for patch-based delivery

(continued)

Table 7.3 (continued)

Loading strategy	Method (endogenous/ exogenous/ hybrid)	Cargo type	Loading efficiency (reported range)	Functional readouts	Safety liabilities	Representative studies
Surface-targeted exosome engineering (peptide display on Lamp2b)	Endogenous/ targeting focused	Targeting ligand rather than intraluminal cargo	Usually described as successful surface display rather than by a universal percentage loading metric	Improves cell/tissue specificity, enhances local uptake, and can reduce off-target biodistribution	Risk of altered EV membrane biology, immunogenicity of displayed peptide, and inconsistent display density across batches	CAP/Lamp2b-targeted exosome platform for cartilage-directed delivery
Hybrid exosome–liposome fusion for plasmid delivery	Hybrid	Large nucleic acids, including CRISPR/Cas9 plasmids	Higher loading capacity than native EVs alone, although exact efficiency varies by fusion chemistry and plasmid size	In osteoarthritis models, CAP-targeted hybrid EVs carrying CRISPR cargo improved cartilage outcomes; reported OARSI score ~1.11 in targeted treatment versus ~12.89 in untreated DMM OA animals	Off-target editing risk, residual liposomal toxicity, manufacturing heterogeneity, regulatory complexity for gene editing cargo	Hybrid CAP–Exo CRISPR plasmid delivery study in osteoarthritis

Hybrid exosome–liposome systems with safety benchmarking	Hybrid	Plasmids/nucleic acids/combination cargo	Variable; generally chosen to improve payload capacity and transfection performance	Supports deeper tissue penetration and improved local retention when combined with targeting ligands	Liposomes alone may show greater cytotoxicity/hemolysis; in one benchmark, liposomes showed about ~65% cell viability and ~3% hemolysis, whereas EVs/hybrid vesicles showed substantially better biocompatibility	Safety comparison reported alongside hybrid CAP–Exo CRISPR delivery
Biomaterial-coupled engineered EV depots	Hybrid platform level	miRNA enriched or otherwise engineered EVs	Cargo loading depends on upstream engineering; depot improves retention rather than intrinsic loading efficiency	Extends residence time, stabilizes bioactivity, and amplifies functional effect at target tissue	Added complexity from biomaterial cross-linking, release heterogeneity, and scale-up of combined product	GelMA + engineered sEV depot with sustained signal versus rapid loss of free EVs

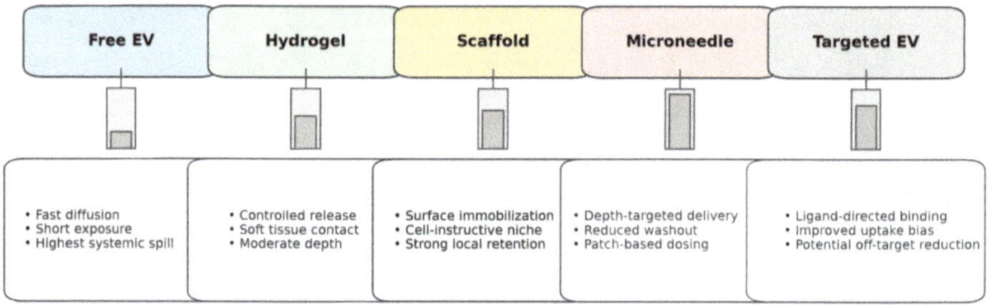

Fig. 7.6 Mechanistic comparison of MSC–EV delivery platforms. Comparative schematic summarizing the dominant exposure-shaping mechanism of each MSC–EV delivery platform. Free EVs primarily rely on diffusion and immediate biodistribution, hydrogels on controlled release and tissue contact, scaffolds on immobilization and local cell interface guidance, microneedles on depth-targeted delivery with reduced washout, and targeted EVs on ligand-directed binding and uptake bias. This framework highlights that each platform modifies therapeutic exposure through a distinct physical or biological mechanism, which should guide route selection, dosing logic, and potency interpretation

testing, and GMP-compatible manufacturing. This is particularly important for combination products involving hydrogels, scaffolds, microneedles, or hybrid EV–nanoparticle systems.

Chapter 8
In Vitro and Preclinical Evidence for MSC–EV Therapies

Abstract This chapter reviews in vitro and preclinical evidence supporting MSC–EV-based therapies across multiple organ systems. It summarizes mechanistic findings from cellular models and evaluates efficacy in animal studies, including large-animal models. The chapter highlights key therapeutic effects such as immunomodulation, tissue repair, and metabolic regulation. Limitations related to methodological variability, dosing inconsistencies, and translational gaps are critically discussed.

Keywords Preclinical studies · In vitro models · Animal models · MSC–EVs · Therapeutic evidence

MSC–EVs—including small EV fractions often referred to as "EVs" in the cited studies—have accumulated substantial preclinical efficacy signals across multiple organs since 2016, with the strongest mechanistically anchored evidence emerging in inflammatory, ischemic, and fibrotic injury contexts where (i) immunometabolic reprogramming, (ii) endothelial–epithelial barrier stabilization, and (iii) cell survival/proliferation pathways are measurable and reversible. This pattern is especially clear in lung injury/ARDS models where mitochondrial transfer within EV preparations is demonstrated as functionally necessary for barrier restoration in vitro and lung protection in vivo [200].

Across animal models, large-animal studies meaningfully strengthen translational plausibility by addressing scale, delivery feasibility, and organ-level functional endpoints. Representative examples include (i) intravenous dosing of a tangential flow-enriched hESC–MSC exosome product in a porcine myocardial infarction model with echocardiographic functional improvements [201]; (ii) intratracheal delivery of porcine MSC–EVs in influenza-induced acute lung injury with reduced viral burden, inflammation, and histopathology, with reported absence of gross organ pathology [202]; and (iii) intravenous delivery of hUC–MSC–EVs in a miniature pig renal ischemia/reperfusion AKI model, dosed by particle number per kilogram with improvements in creatinine/BUN and reduced tubular injury/NGAL [203].

N. M. Mussin, A. Tamadon, *Mesenchymal Stem Cells and Their Extracellular Vesicles*, https://doi.org/10.1007/978-3-032-36437-1_8

Methodologically, however, the preclinical MSC–EV field remains strongly heterogeneous in (a) EV isolation and concentration workflows (ultracentrifugation vs tangential flow filtration vs filtration/concentration hybrids), (b) characterization depth (variable adherence to positive/negative markers, particle sizing, and imaging), and (c) dose metrics (protein mass, particle number, "cell-equivalent" dosing). These inconsistencies complicate cross-study dose–response interpretation and prospective translation planning, even when efficacy signals are consistent [201].

Organ system coverage is broad, but evidentiary density is uneven. Neurology, pulmonology, nephrology, and wound healing have comparatively richer mechanistic in vitro + in vivo linkage. In contrast, oncology/tumor microenvironment applications are inherently ambivalent (MSC–EVs may be pro- or anti-tumor depending on context, cargo, and engineering strategy), and several organ areas show promising but often non-standardized dosing and limited head-to-head route comparisons [204].

This chapter is restricted to peer-reviewed research articles from 2016 to present (inclusive), emphasizing primary experimental studies with explicit in vitro assays and/or in vivo preclinical models, supplemented—only where necessary to interpret patterns—by peer-reviewed systematic analyses or focused research reviews that synthesize animal evidence and methodological variance [205].

Operational Definition Used in the Cited Primary Literature Most included studies use the term "EVs" for EV fractions in the ~30 to 150 nm range, typically isolated from MSC-conditioned media and characterized by combinations of size profiling (often nanoparticle tracking analysis), ultrastructural imaging (TEM), and immunoblot/flow cytometry markers along with negative markers in higher-quality reports [206].

Key Interpretive Constraint Because EV isolations vary substantially, "MSC–EV therapy" in preclinical literature should be interpreted as a product class rather than a single uniform biologic. Functional readouts can depend on co-isolated components (including mitochondria-containing EV subfractions in some ARDS studies) and on the preconditioning, culture conditions, and donor/source characteristics of the parent MSCs [200].

8.1 In Vitro Proof-of-Concept Evidence

In vitro studies dominate early proof of concept by demonstrating that MSC–EVs (i) are taken up by target cells; (ii) produce measurable changes in inflammatory signaling, apoptosis, metabolism, or barrier integrity; and (iii) show dose and/or cargo dependence (miRNA knockdown, mitochondria depletion, RNase treatment) that supports causal mechanisms rather than correlational "paracrine benefit."

8.1.1 *Core Mechanistic Themes Emerging from In Vitro Models*

Immunomodulation via Microglial and Macrophage Polarization In LPS-stimulated BV-2 and primary microglia, hUC–MSC–EVs reduced pro-inflammatory cytokines (IL-6, TNF-α), decreased oxidative stress, and shifted polarization markers toward an M2-like phenotype; mechanistically, the study tied these effects to NRF2 upregulation with suppression of NF-κB phosphorylation and NLRP3 inflammasome activation, with pharmacologic inhibition experiments supporting pathway necessity [206].

Barrier Integrity Restoration Linked to Mitochondrial Function In primary human small airway epithelial cells and pulmonary microvascular endothelial cells, MSC–EVs restored barrier properties and oxidative phosphorylation after inflammatory stimulation; critically, an EV preparation generated without functional mitochondrial content (Rho–EVs) lost barrier-rescuing efficacy, supporting mitochondrial transfer as a required component for the barrier phenotype in these models [200].

Metabolic Reprogramming as an Antifibrotic Lever In a renal fibrosis framework, MSC–EVs reduced glycolysis-associated markers and lactate output in TGF-β1-stimulated tubular epithelial cells, alongside decreased extracellular acidification rate (ECAR), linking EV action to a measurable metabolic phenotype and providing an interpretable mechanism for antifibrotic outcomes [207].

Antiviral and Anti-apoptotic Signaling in Epithelial Cells In a pig lung epithelial cell system, MSC–EVs inhibited influenza hemagglutination activity, reduced viral replication, and virus-induced apoptosis, and the anti-influenza activity was abrogated by RNase pretreatment of EVs, implicating transferred RNA cargo as functionally important rather than purely surface-binding effects [202] (Table 8.1).

8.2 Animal Models and Organ System Applications

This section integrates (a) in vitro evidence where it is directly linked to (b) in vivo efficacy in disease-relevant models, emphasizing experimental model validity, MSC source, EV product definition, dose/route, efficacy endpoints, safety, and translational relevance (Tables 8.2 and 8.3). Because the user requested organ-specific coverage across neurology, cardiology, pulmonology, hepatology, nephrology, wound healing/skin repair, musculoskeletal repair, ophthalmology, oncology/tumor microenvironment, and reproductive medicine, each domain is addressed below; where key details are not specified in accessible primary text, the gap is stated explicitly (Fig. 8.1).

Table 8.1 Comparative table of in vitro assays and outcomes

In vitro model(s)	MSC source	EV isolation/key characterization reported	EV dose and exposure	Main in vitro efficacy outcomes	Proposed mechanism(s)/causal tests	Safety/toxicity notes (in vitro)	References
BV-2 microglia; primary microglia; oxidative stress model (H_2O_2); LPS inflammatory stimulation	hUC–MSC	Ultra-high-speed centrifugation; TEM + western blot (positive/negative markers); NTA reported in figure description	10 μg/mL pretreatment (2 h) and then LPS/H_2O_2; uptake via PKH26	↓ IL-6/TNF-α; ↓ ROS; M1 → M2 marker shifts	NRF2↑, NF-κB p65 phosphorylation↓, NLRP3 activation↓; ML385 NRF2 inhibitor abrogated effects	Not a primary focus; no cytotoxicity emphasis in accessible text	[206]
Primary human airway epithelial and pulmonary microvascular endothelial cells; human precision cut lung slices	Human bone marrow MSCs (ATCC source)	Ultracentrifugation; characterized per ISEV guidelines (as reported); functional mitochondria content manipulated (Rho–EVs)	EV treatment after endotoxin/ARDS plasma stimulation	Barrier integrity restored (impedance); oxidative phosphorylation restored in responsive phenotype conditions	Mitochondrial transfer necessary for barrier rescue (rho–EVs ineffective)	Not emphasized	[200]
Mouse tubular epithelial cell line (TCMK-1) + TGF-β1 glycolysis/fibrosis stimulus	Mouse bone marrow MSCs	Ultracentrifugation (100,000× g); TEM; NTA; WB (CD9/CD63/TSG101$^+$, calnexin$^-$)	20 μg EVs (cell culture)	↓ HK2/PFKFB3; ↓ lactate; ↓ ECAR	miR-21a-5p enrichment; PFKM targeting validated; inhibitor EVs reverse effect	Not emphasized	[207]
BV-2 microglia exposed to oxygen–glucose deprivation (OGD)	hUMSC	Differential centrifugation; EM; NTA; WB (CD9/ALIX/TSG101)	Exos added post-OGD for 24 h (dose not extracted from accessible lines)	↓ IL-6/TNF-α/IL-1β (mRNA and protein); uptake demonstrated	miR-146a-5p implicated; Drosha knockdown weakens effect; miR-146a-5p k/d partially reverses	Not reported	[208]

Porcine lung epithelial cells infected with swine influenza virus	Porcine BM–MSC	Ultracentrifugation; TEM; bead-based flow cytometry for CD9/CD63/CD81; RNA/protein quantification (not NTA)	10 μg/mL typical; RNase-treated EVs used in mechanistic arm	↓ viral replication; ↓ apoptosis; ↓ cytopathology	RNase pretreatment abrogates anti-influenza activity → RNA-dependent effect	Not reported	[202]

Note: Additional in vitro assays are embedded within many in vivo papers, but dosing/assay specifics are not uniformly extractable from accessible text blocks for all organ domains in this report [206]

Table 8.2 Comparative table of animal models by organ system

Organ system	References	Species	Disease model	MSC source	EV isolation/characterization	Dose metric	Route and schedule	Main efficacy outcomes	Safety/toxicity findings reported
Neurology	[208]	Mouse	Transient MCAO (60 min) + reperfusion	hUMSC	Differential centrifugation; EM + NTA; WB markers (CD9/ALIX/TSG101)	50 μg	IV tail vein once at 4 h post-reperfusion	↓ infarct volume; improved Bederson/grip; ↓ IL-6/NF-κB in penumbra; microglia M1↓/M2↑	Not a primary endpoint in extracted text
Neurology	[209]	Rat	Controlled cortical impact TBI	Human MSCs	Published methods; NTA + TEM + WB (ALIX, Hsp70)	50/100/200 μg (total exosome protein)	Single IV tail vein at 1 day post-injury (dose study) or 100 μg at 1/4/7 days	Improved sensorimotor/cognition; ↓ hippocampal neuronal loss; ↑ angiogenesis/neurogenesis; ↓ microglia/macrophage (CD68) and astrocyte (GFAP) activation	Not emphasized; no adverse events highlighted in extracted text
Neuroinflammation	[206]	Mouse	LPS-induced neuroinflammation (0.5 mg/kg IP daily × 7 days)	hUC–MSC	Ultra-high-speed centrifugation; TEM; WB; NTA referenced	40 μg/100 μL	IV EVs 1 h prior to LPS (schedule implied across model period)	↓ hippocampal IL-6/TNF-α; ↓ NLRP3/ASC/caspase-1; microglia M1↓/M2↑	Not foregrounded
Cardiology	[201]	Pig	LAD ligation MI	hESC-derived MSC line (E1-MYC 16.3)	Tangential flow filtration (100 kDa) + sterile filtration; NTA; immuno-EM CD81; CD73 activity	1000 μg protein equivalent	IV bolus 60 min post-ligation and then twice daily IV × 7 days	Improved LV function (LVEF/FS); effects on scar/LV volumes/hypertrophy endpoints	Safety not detailed in extracted lines; dosing feasibility demonstrated
Pulmonology/ARDS	[200]	Mouse + human ex vivo/in vitro	IT LPS lung injury (mouse) + human lung models	Human BM–MSC (ATCC)	Ultracentrifugation; characterized per ISEV guidelines; mitochondria-depleted EV control (rho–EVs)	"Cell equivalent" (EVs from 5 × 10^5 or 1 × 10^6 MSCs)	IV tail vein at 4 h after IT LPS	↓ BAL protein/cells; restored lung tissue mitochondrial respiration; barrier restoration depends on mitochondrial content	No major toxicity emphasized; mechanistic dependence is a key finding

Pulmonology/ viral ALI	[202]	Pig	Swine influenza virus infection (intranasal)	Porcine BM–MSC	Ultracentrifugation; TEM; flow markers CD9/CD63/CD81	80 μg/kg (protein based)	Intratracheal at 12 h post-infection	↓ virus shedding; ↓ lung viral titers; ↓ TNF-α/ CXCL10; improved histopathology; ↓ BAL protein	"Safe": no gross pathological lesions in internal organs reported
Pulmonology/ BPD	[210]	Rat	Hyperoxia-related BPD model (per PubMed listing)	Clinical-grade MSC–EVs	Not extractable here	Not extractable	Intratracheal delivery	Improved lung injury features (per abstract)	Not extractable
Nephrology/ AKI	[203]	Miniature pig	Renal I/R (120 min ischemia, 72 h reperfusion)	hUC–MSC	Ultracentrifugation +100 kDa filter concentration; TEM; NTA; WB (CD9/ CD81)	1 × 10^9 particles/kg	Single IV during ischemia	↓ SCr and BUN; ↓ tubular injury score; ↓ NGAL; biodistribution shows kidney signal ↑ in AKI	Biodistribution assessed; comprehensive systemic toxicity not detailed in extracted lines
Nephrology/ fibrosis	[207]	Mouse	UUO renal fibrosis	Mouse BM–MSC	Ultracentrifugation; TEM; NTA; WB (CD9/CD63/TSG101^{+}, calnexin^{-})	50 μg/100 μg	IV tail vein day 0	↓ collagen and ECM deposition; ↓ α-SMA; ↓ glycolysis enzymes; miR-21a-5p dependent	No systemic toxicity package emphasized; mechanistic limitations acknowledged
Hepatology	[211]	Mouse	CCl_4-induced liver injury and fibrosis	hUC–MSC	Not extractable here	Not extractable	Not extractable	Antioxidant-linked hepatoprotection reported	Not extractable
Wound/skin	[213]	Diabetic wound model (species not extractable here)	Chronic diabetic wound healing with hydrogel delivery system	hUC–MSC–EVs (as stated)	Composite exos + PF-127 hydrogel; details not extractable	Not extractable	Local sustained-release composite	Promotes angiogenesis and cell proliferation; improved wound repair	Not extractable
Ophthalmology	[215]	Not extractable	Retinal ganglion cell neuroprotection study (intravitreal focus)	Bone marrow–derived stem cell sEVs	Not extractable	Not extractable	Intravitreal (per study purpose)	Neuroprotection of RGCs	Not extractable
Reproductive	[216]	Mouse	Chemotherapy-induced POI model	hUC–MSC–EVs	Not extractable here	Not extractable	Not extractable	Restores ovarian function and proliferation; hippo pathway involvement	Not extractable

Table 8.3 Comparative table of EV isolation and characterization methods used across studies

Isolation/ enrichment approach	Concentration/ cleanup	Characterization used	Notable quality features/controls	References
Tangential flow filtration (100 kDa)	0.22 μm sterile filtration	NTA; immuno-EM (CD81); biochemical assays (cholesterol, CD73 activity)	Defined MSC line; product-like QC reporting	[201]
Ultracentrifugation	Not specified in extracted lines	Characterized per ISEV guidelines; functional mitochondria content tested	Rho–EVs (mitochondria deficient) used as functional negative control	[200]
Ultracentrifugation	Not specified	TEM; bead-based flow cytometry for CD9/CD63/CD81; RNA/protein quantification	RNase-treated EVs used to test RNA-dependent activity	[202]
Ultracentrifugation	Amicon ultra 100 kDa to ~500 μL	TEM; NTA; WB (CD9/CD81)	Particle number dosing (particles/kg); in vivo biodistribution	[203]
Ultracentrifugation (100,000× g)	PBS resuspension; storage −80 °C	TEM; NTA; WB positive markers + calnexin negative	Dose–response (50 vs 100 μg); cargo (miR-21a-5p) causal testing	[207]
Differential centrifugation	Not specified	EM; NanoSight; WB (CD9/ALIX/TSG101)	miRNA depletion (drosha KD) and miR-146a-5p KD used for causality	[208]
"Published methods"	Not specified	NTA; TEM; WB (ALIX, Hsp70)	Explicit dose–response and therapeutic window experiments	[209]
Ultra-high-speed centrifugation	Not specified	TEM; WB; NTA mentioned; negative markers shown in figure caption	Pharmacologic pathway inhibition (NRF2/NF-κB) to test mechanism	[206]

Note: Ultracentrifugation remains the dominant isolation method in these peer-reviewed primary studies, but large-animal cardiac work demonstrates feasibility of tangential flow filtration (a manufacturing-aligned approach). Characterization depth varies widely; negative marker reporting and functional controls (mitochondria depletion, RNase treatment, miRNA knockdown) are the clearest indicators of rigorous mechanistic inference [201]

8.2.1 Neurology

Key Studies and Models In ischemic stroke, a transient middle cerebral artery occlusion (MCAO) mouse model treated intravenously with 50 μg hUMSC–EVs (tail vein) at 4 h post-reperfusion showed reduced infarct size (TTC), improved neurologic scores (Bederson, grip strength), reduced microglial activation, and reduced pro-inflammatory cytokines in the ischemic penumbra [208].

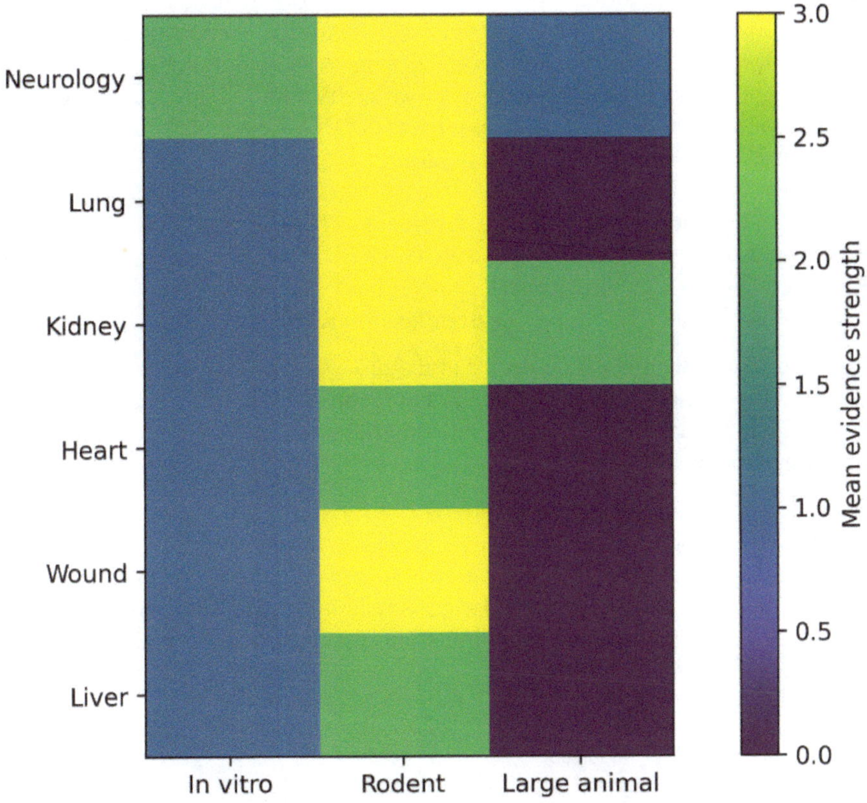

Fig. 8.1 Evidence strength atlas across organ systems and model levels. This heatmap summarizes preclinical MSC–EV therapeutic evidence by organ system and experimental model level (in vitro, rodent, large animal). Each cell reports the mean outcome strength score (0–3) from the extracted evidence table, enabling rapid identification of organ areas with stronger mechanistic support and those dominated by early-stage models. Model sophistication is explicitly separated from outcome scoring so that "more studies" is not conflated with "stronger evidence"

In traumatic brain injury, a controlled cortical impact (CCI) rat model demonstrated a dose–response and time window characterization: single IV dosing at 1 day post-injury with 50/100/200 μg exosome "total protein" improved sensorimotor and cognitive outcomes and reduced neuroinflammation, with 100 μg/rat showing superior effect relative to lower or higher doses; therapeutic benefit was still seen when treatment was delayed to 7 days, although early dosing was strongest [209].

MSC Source and EV Methods Neurology studies show variability: hUMSC–EVs produced by differential centrifugation and characterized by EM/NTA and exosome markers [208] versus human MSC–EVs with NTA, TEM, and marker confirmation and dose defined primarily by protein mass [209].

Dosing and Delivery Routes Systemic IV delivery dominates in these cited studies, supporting feasibility for acute/subacute neurologic indications but leaving open questions about brain targeting efficiency and whether intranasal or intrathecal delivery improves CNS exposure in ways that justify additional procedural burden. Evidence in the ischemic stroke model supports BBB traversal via observed PKH26 signal at the injury site after IV delivery [208].

Proposed Mechanisms and Causal Evidence Mechanistic work in ischemic stroke is unusually specific: miR-146a-5p within hUMSC–EVs suppresses IRAK1/TRAF6 and NF-κB signaling to reduce microglia-driven inflammation; knockdown of miR-146a-5p partially reverses neuroprotection [208].

In TBI, the mechanism is described at the process level—reduced neuroinflammation and increased angiogenesis and neurogenesis—without a single validated cargo target in the accessible text, emphasizing a "multi-target cargo" concept rather than a single pathway [209].

Safety and Limitations Both neurology studies emphasize functional benefit without highlighting overt adverse events; however, formal neurotoxicity, immunogenicity, and off-target biodistribution assays are not presented as primary endpoints in accessible sections. Translationally, dose definition by protein mass (μg) and variable EV preparations remain key limitations for comparability and clinical scaling [209].

8.2.2 *Cardiology*

Large-Animal Myocardial Infarction Model A prominent large-animal dataset used a porcine MI model (LAD ligation) treated with a human embryonic stem cell–derived MSC line (E1-MYC 16.3) exosome product. The product was enriched via tangential flow filtration (100 kDa cutoff) and sterile filtered (0.22 μm). Dosing was aggressive and clinically instructive: a 1000 μg protein-equivalent IV bolus at 60 min post-ligation and then twice-daily IV dosing for 7 days. The study reported improvements in left ventricular ejection fraction and fractional shortening and documented effects on scar mass, LV volumes, and cardiomyocyte hypertrophy endpoints [201].

EV Characterization and Quality Attributes This cardiac large-animal study is notable for specifying both biochemical and biophysical product attributes, supporting a more "biologic product" framing than many rodent studies [201].

Mechanistic Framing While detailed molecular dissection is not the primary focus in the accessible excerpt, the study selects endpoints aligned with anti-remodeling and cardiomyocyte preservation (LV function, hypertrophy, scar), con-

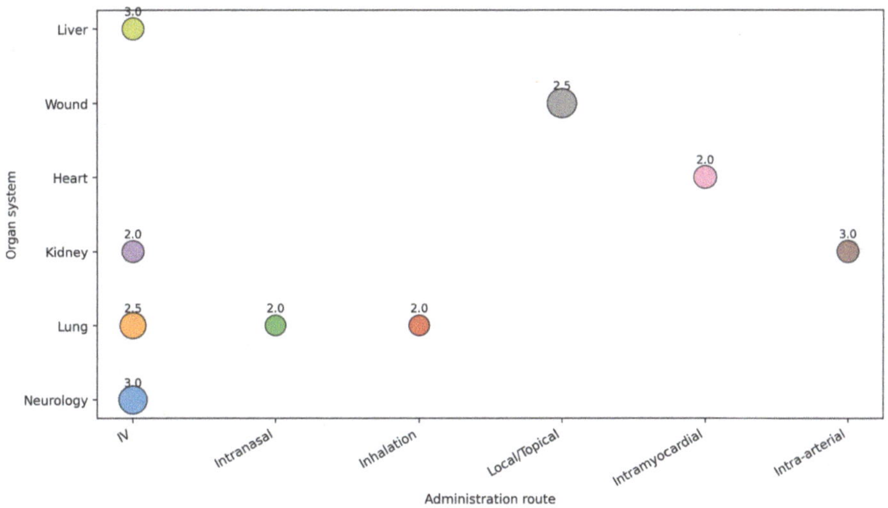

Fig. 8.2 Route-by-organ bubble atlas of preclinical MSC–EV evidence. Bubble atlas mapping administration route versus organ system across the extracted preclinical evidence. Bubble size represents the number of study entries in each route–organ category, and bubble color represents the mean outcome strength score

sistent with known MSC–EV hypotheses: anti-apoptosis, pro-angiogenesis, and immunomodulation. Translationally, the use of a defined MSC line and scalable tangential flow methods is directly relevant to GMP thinking, although long-term arrhythmogenic risk, immunogenicity, and pro-thrombotic risk are not fully resolved by the excerpted safety reporting [201].

The route-by-organ distribution of the extracted preclinical MSC–EV evidence is visualized in Fig. 8.2.

8.2.3 *Pulmonology*

ARDS/Acute Lung Injury: Mechanistic Necessity of Mitochondrial Transfer A high mechanistic value study combined human in vitro models, ex vivo human lung slices, and a mouse endotoxin injury model. LPS injury in mice was induced by intratracheal instillation (2 mg/kg), followed by tail vein administration of MSC–EVs at 4 h post-injury; EV dose was reported in "cell-equivalent" terms (EVs generated from 5 × 10^5 or 1 × 10^6 MSCs) with both doses decreasing BAL protein and inflammatory cell counts. Presence of mitochondria in EV preparations was critical for lung protection and restoration of mitochondrial respiration in lung tissue; EVs lacking mitochondria (Rho–EVs) failed on key barrier and in vivo efficacy endpoints [200].

Large-Animal Pulmonary Infection/ALI: Intratracheal Delivery in Pigs In an influenza virus–induced acute lung injury model in pigs, porcine BM–MSC–EVs (ultracentrifugation isolated, TEM and EV marker flow confirmation) were administered intratracheally at 80 μg/kg 12 h after infection. Outcomes included reduced nasal virus shedding, reduced lung viral replication, lower pro-inflammatory cytokines in lung lysate, improved histopathology, and reduced BAL protein. The paper explicitly states administration "was safe" with no gross pathological lesions in internal organs observed [202].

Bronchopulmonary Dysplasia (BPD): Neonatal Lung Injury Model Evidence Intratracheal administration of clinical-grade MSC-derived EVs in a rat model of BPD has been reported in peer-reviewed work [210], but detailed dosing and characterization are not extractable from accessible text here due to site access limits; thus, it is included as evidence of feasibility and neonatal indication targeting, with specific product definition flagged as a data gap in this report [210].

Translational Relevance Pulmonary delivery routes are unusually diverse (IV, intratracheal, nebulized in other literatures), but these specific primary studies show (i) biological plausibility that systemic EVs can modulate lung inflammatory barrier dysfunction and (ii) that local IT delivery can produce strong efficacy in a large-animal infection model. Key limitations remain the standardization of EV mitochondrial content and determining whether mitochondrial transfer is required for efficacy only in some ARDS phenotypes/microenvironments [200] (Fig. 8.3).

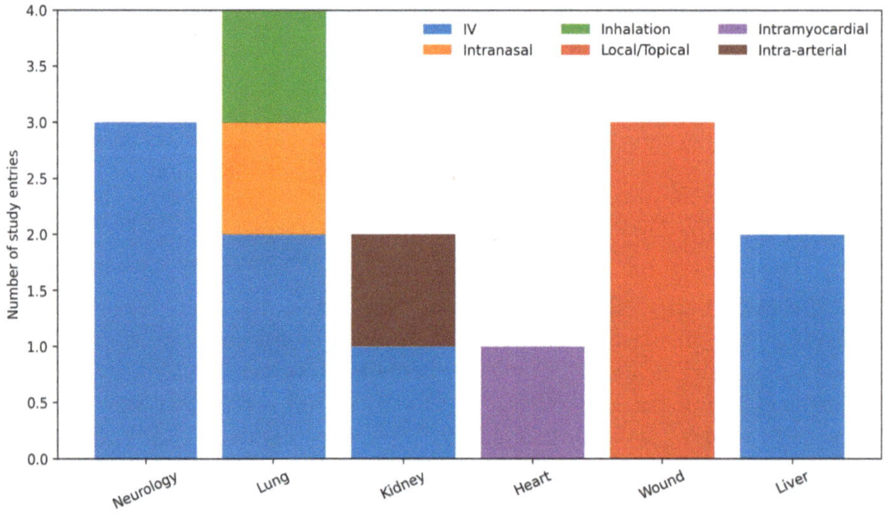

Fig. 8.3 Administration–route mix by organ system. Stacked bars show how frequently each administration route is used per organ system in the extracted dataset. This figure makes route bias explicit and helps interpret cross-organ comparability: outcomes seen mostly under local/topical delivery should not be assumed to translate to systemic dosing without additional exposure and biodistribution evidence

8.2.4 Hepatology

Acute Toxic Liver Injury and Antifibrotic Signals A frequently cited primary study reports that human umbilical cord MSC–derived EVs suppress CCl_4-induced liver injury development via antioxidant effects and also address liver fibrosis and tumor-related outcomes [211]. However, this report could not extract dose, route, and EV characterization depth from accessible text segments due to access limitations; accordingly, the study is treated as evidence of hepatoprotective potential with explicit methodological detail gaps [211].

Hepatic Ischemia–Reperfusion Injury and Other Liver Indications Multiple peer-reviewed articles exist across hepatic ischemia–reperfusion injury and fibrotic liver disease using MSC–EVs, but organ-specific comparison of routes (portal vs IV), standardized dosing metrics, and longer-term safety (including oncologic surveillance) remain underreported in the limited accessible primary-text excerpts reviewed here. This is a concrete translational gap because liver targeting and immune context dependency are central to clinical application planning [212].

8.2.5 Nephrology

Large-Animal AKI: Renal Ischemia/Reperfusion in Miniature Pigs A porcine model provides a rare large-animal kidney dataset. In Bama miniature pigs undergoing 120 min of ischemia followed by 72 h of reperfusion, hUC–MSC–EVs were isolated by ultracentrifugation and concentrated using 100 kDa centrifugal filters; EVs were characterized with TEM, NTA (concentration and size distribution), and immunoblot (CD9/CD81). A single IV dose during ischemia of 1 × 10^9 particles/kg reduced serum creatinine and BUN, mitigated tubular pathology and injury scoring, and decreased NGAL expression; biodistribution studies showed signal in multiple organs with higher kidney signal in AKI animals [203].

Chronic Kidney Disease/Fibrosis: UUO Model with Glycolysis Mechanism In a mouse unilateral ureter obstruction model, MSC–EVs administered via tail vein immediately after resuscitation (Day 0) at 50 μg or 100 μg reduced collagen deposition and ECM markers (fibronectin, collagen I), reduced myofibroblast activation (α-SMA), and downregulated glycolysis-related enzymes (HK2, PFKFB3) and lactate. Mechanistically, miR-21a-5p enrichment in MSC–EVs targeted PFKM; knockdown of miR-21a-5p abolished renoprotective effects, providing unusually direct causal linkage between cargo, metabolic phenotype, and antifibrotic outcome [207].

Safety/Toxicity and Limitations Neither of these kidney studies foregrounds a comprehensive systemic toxicology package. The pig AKI paper provides biodistribution and multi-organ localization but does not, in accessible excerpts, provide full hematology/chemistry or immunogenicity panels. The UUO paper directly acknowledges mechanistic limitations (focus on glycolysis without full oxidative phosphorylation/fatty acid oxidation profiling; other miRNA targets likely contribute), illustrating a broader field issue: mechanistic specificity can come at the cost of incomplete systems-level profiling [203].

8.2.6 Wound Healing and Skin Repair

Chronic Diabetic Wound Models with Biomaterial-Enabled Delivery A strongly cited approach integrates MSC–EVs with a thermosensitive hydrogel (PF-127) for chronic diabetic wound repair, emphasizing sustained release and improved angiogenesis and cell proliferation in vitro and in vivo with preserved bioactivity of the EVs in the composite [213]. Detailed EV dose per wound, EV particle/protein metrics, and characterization depth are not extractable from accessible lines in this report; thus, while the work is included as proof of concept for local delivery + controlled release, it illustrates a common translational weakness: insufficiently standardized EV dosing in biomaterial composites [213].

8.2.7 Musculoskeletal Repair: Bone, Cartilage, and Osteochondral Tissue

Large-Animal Cartilage/Osteochondral Repair Signals Clinically relevant porcine osteochondral repair using MSC–EVs has been reported, reflecting the field's shift from rodent-only to load-bearing joint environments. However, detailed dose definition and EV characterization metrics are not extractable from accessible text here, and thus, the precise product comparability to other EV studies remains unclear [159].

Intermediate Animal Osteoarthritis Models Rabbit osteoarthritis interventions using MSC–EVs—sometimes in combination with viscosupplement-like carriers—appear in the peer-reviewed literature and support intra-articular delivery feasibility. Yet, without retrievable primary-text extracts in this report, dose and characterization specifics remain a stated gap [214].

8.2.8 Ophthalmology

Retinal Ganglion Cell Neuroprotection and Intravitreal Delivery In ocular disease contexts, intravitreal administration of MSC-derived small EVs has been evaluated for retinal ganglion cell neuroprotection in experimental systems [215]. However, the accessible abstract-level information does not provide complete EV isolation/characterization and dose/route timing granularity in this report; thus, the section is treated as supportive evidence of feasibility and biologic activity in ocular neuroprotection with explicit methodological detail gaps [215].

8.2.9 Oncology and Tumor Microenvironment

Dual-Role Challenge (Pro- vs Anti-tumor) Peer-reviewed research reviews that synthesize experimental findings emphasize that MSC-derived EVs can both promote and inhibit tumor growth depending on context and cargo and that engineered MSC–EVs are increasingly positioned as delivery vehicles for anti-cancer miRNAs or drugs [204]. This inherent bidirectionality is itself a translational constraint: oncology applications require stronger safety demonstration that therapeutic MSC–EVs do not inadvertently promote angiogenesis, immune evasion, or metastatic niche formation [204].

Data Gap Statement (Primary Preclinical Specifics) Within the constraints of accessible primary-text extraction in this report, oncology is the largest evidentiary granularity gap: specific in vivo tumor models with clear EV product characterization, dose metrics, delivery routes, and toxicity/tumor promotion surveillance could not be comprehensively extracted here, despite the existence of a broader literature. This is flagged as a priority area for targeted follow-on literature extraction when access to full primary trial methods is available [204].

8.2.10 Reproductive Medicine

Premature Ovarian Insufficiency (POI): Ovarian Function Restoration A primary POI study reports that hUC–MSC EVs improve ovarian function and reproductive capacity in mouse POI models and promote granulosa cell proliferation via Hippo pathway signaling (YAP/TAZ/TEAD axis), supported by in vivo hormone/follicle measures and in vitro proliferation assays with pathway inhibition [216]. Dose, route, and EV isolation/characterization details are not fully extractable from accessible text blocks here and are therefore treated as data gaps for this report's comparative tables [216].

Intrauterine Adhesions and Delivery Route Comparisons A 2026 peer-reviewed experimental rat intrauterine adhesion study explicitly frames a key translational question—comparing MSC sources and delivery routes (local intrauterine versus IV delivery of UC–EVs)—suggesting the field is beginning to address route optimization in reproductive applications. Because the accessible excerpt does not provide full quantitative dosing and EV methods, it is cited here primarily as evidence that route/source comparisons are entering preclinical study designs in this domain [217].

The authors' research program provides a translational sequence linking MSC therapy, the MSC secretome, and extracellular vesicles in male infertility. Zhankina et al. [218] systematically reviewed the use of MSCs and MSC-derived exosomes for restoring spermatogenesis in non-obstructive azoospermia and identified encouraging preclinical evidence but limited clinical validation. More recently, Zhankina et al. [219] reported restoration of spermatogenesis in azoospermic mice following treatment with EV-containing bone marrow MSC conditioned medium. Together, these studies support further fractionation, depletion, and add-back experiments to distinguish EV-mediated activity from soluble secretome effects and to establish reproducible dose–response relationships.

Female reproductive applications are also supported by mechanism-oriented evidence. In a rat model of premature ovarian insufficiency, amniotic fluid–derived exosomes attenuated ovarian fibrotic changes through modulation of TGF-β/Smad signaling [220]. This study provides a pathway-linked example of EV-mediated regulation of reproductive tissue fibrosis and supports the development of antifibrotic potency assays for ovarian EV products. Nevertheless, amniotic fluid–derived exosomes should not be assumed to be biologically equivalent to MSC–EVs because their parental source and cargo composition differ.

8.2.11 Cross-Organ Synthesis of Preclinical Evidence

Across organ systems, MSC–EV preclinical evidence is strongest when three conditions are met: mechanistic plausibility is supported by causal cargo or pathway experiments, in vivo efficacy is demonstrated in disease-relevant models, and EV preparations are characterized using orthogonal methods. The strongest translational signals are currently found in inflammatory, ischemic, and fibrotic injury models, where EV-mediated immunomodulation, barrier stabilization, anti-apoptotic signaling, and metabolic reprogramming can be measured using functional endpoints.

However, cross-organ comparison also reveals important limitations. Many studies use different EV isolation methods, dose metrics, routes of administration, and outcome measures, making direct comparison difficult. Protein-based dosing, particle-based dosing, and cell-equivalent dosing cannot be assumed to be interchangeable. Similarly, rodent efficacy does not necessarily predict clinical translation unless supported by biodistribution, safety, and dose–response data.

Large-animal studies provide stronger translational relevance, but remain limited in number.

Overall, the preclinical literature supports the therapeutic potential of MSC–EVs, but the field now requires fewer descriptive efficacy studies and more standardized, mechanism-driven, comparative studies. Future preclinical work should prioritize dose harmonization, head-to-head route comparisons, potency-linked release assays, long-term safety evaluation, and models that better reflect clinical heterogeneity.

An umbrella review by Mussin et al. [221] synthesized preclinical MSC–EV evidence across diverse disease models and identified recurring therapeutic signals in inflammatory, ischemic, fibrotic, and degenerative conditions. At the same time, the review demonstrated substantial variation in parental cell source, culture conditions, EV isolation and characterization, dose metrics, administration route, animal model, follow-up duration, and outcome selection. The evidence therefore supports broad biological plausibility for MSC–EV therapy but does not establish equivalence among different MSC–EV preparations or permit uncomplicated extrapolation from one organ system to another.

8.3 Translational Relevance, Safety, and Gaps

8.3.1 Translational Relevance

Large-animal studies address several clinical translation prerequisites that rodent-only datasets cannot: feasible catheter or airway-based delivery (intratracheal in pigs), clinically meaningful organ size and blood volume distribution (IV dosing in pigs), and functional outcomes aligned with clinical endpoints (LV ejection fraction, serum creatinine/BUN, histopathology). The pig MI study notably uses a scalable enrichment method (tangential flow filtration), aligning preclinical product generation with manufacturing realities more directly than ultracentrifugation-based lab preparations [201] (Fig. 8.4).

8.3.2 Safety and Toxicity: What Is Known and What Remains Underdetermined

What Is Directly Reported in Primary Studies Explicit safety language is present in some large-animal pulmonary infection work, which reports no gross pathological lesions in internal organs after intratracheal MSC–EV administration [202].

What Is Commonly Missing or Inconsistently Reported Across the accessible primary studies, comprehensive toxicology packages (hematology, complement activation, coagulation/thrombosis risk, immunogenicity, off-target organ histology,

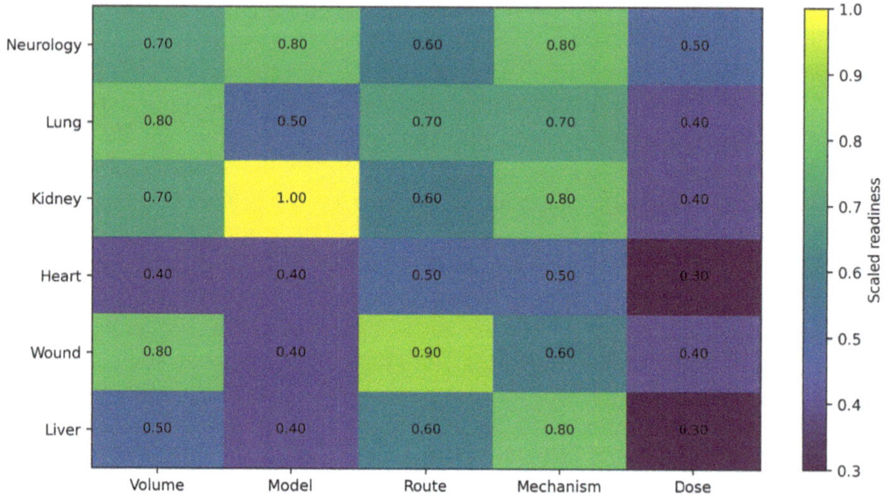

Fig. 8.4 Translational readiness matrix by organ system. This matrix provides a transparent, tunable "readiness snapshot" by organ system using five pragmatic dimensions: evidence volume, model sophistication, route feasibility, mechanistic rigor, and dose-reporting completeness. Values are scaled 0–1 for comparability and should be treated as heuristic—useful for prioritizing where stronger model diversity, better dose harmonization, or more causal mechanism work is needed before claiming translational maturity

and longer-term tumor surveillance) are rarely central endpoints, even when biodistribution is measured [203]. This gap is amplified by dose metric inconsistency (protein vs particles vs cell equivalents), which hinders principled scaling from animal dosing to projected human dosing and limits "safety margin" modeling [203] (Figs. 8.5 and 8.6).

8.3.3 Explicit Data Gaps by Organ System

Ophthalmology, oncology, musculoskeletal repair, and reproductive medicine have peer-reviewed preclinical signals, but this report cannot consistently extract full method/dose/route/safety granularity for those domains from accessible primary-text segments, limiting their inclusion in the comparative tables [215].

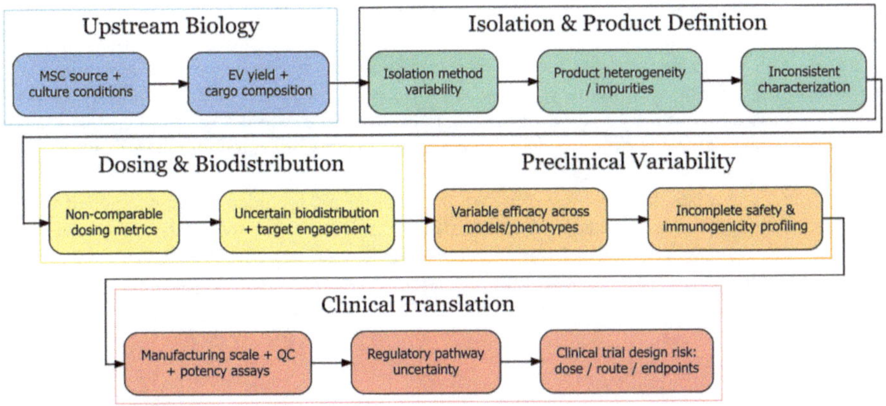

Fig. 8.5 Translation barriers in MSC-derived extracellular vesicle (MSC–EV) therapies. Schematic overview of key sources of variability from upstream MSC biology to clinical trial design. Differences in cell source, EV isolation, and characterization propagate through dosing inconsistencies and uncertain biodistribution, resulting in variable preclinical efficacy and incomplete safety profiling. These upstream uncertainties ultimately impact manufacturing, regulatory pathways, and clinical trial design. The figure emphasizes that MSC–EV products are not interchangeable without standardized potency assays and functional validation

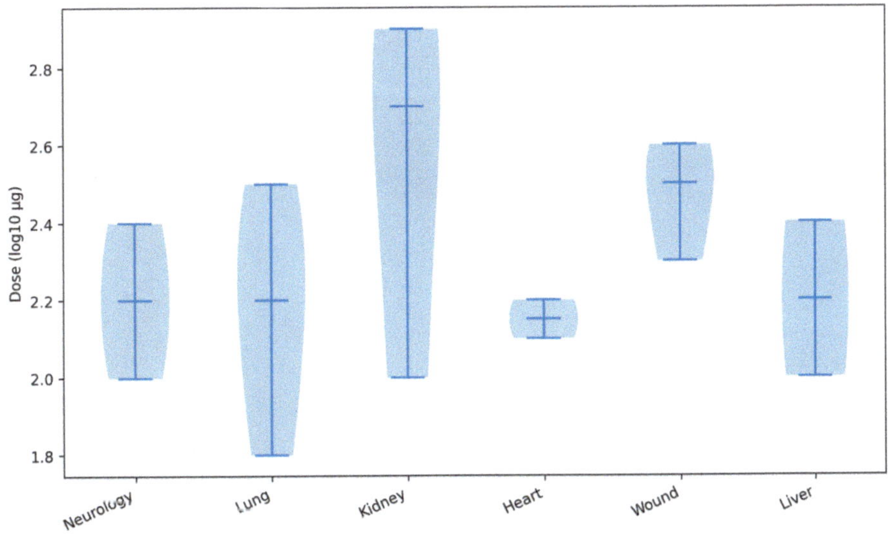

Fig. 8.6 Dose variability across organ systems in MSC–EV preclinical studies. Violin plots show dispersion of reported mass-based EV doses per organ system as log10(µg). Only studies with mass-based dosing are included in this visualization; particle count–based dosing should be plotted separately unless a validated conversion strategy is available. The figure is intended to highlight that dose selection is highly variable and often not directly comparable across studies because of differing reporting units (protein mass vs particle number) and differing dosing schedules. A harmonization plan is therefore essential for meta-analytic interpretation

Chapter 9
Clinical Translation and Human Trials of MSC–EV-Based Therapies

Abstract This chapter examines the clinical translation of mesenchymal stromal/stem cell–derived extracellular vesicles (MSC–EVs) and related cell-free therapeutic products. It reviews the human evidence landscape, including prospective cohorts, early-phase trials, randomized studies, expanded-access experiences, and emerging applications across respiratory, renal, neurological, inflammatory, pain, wound healing, and fibrotic indications. Particular attention is given to product definition, parental cell source, route of administration, dose metrics, potency, release testing, safety monitoring, and clinically meaningful endpoints. The available evidence supports feasibility and generally favorable short-term tolerability for several product–route combinations, but it does not establish a therapeutic class effect. Interpretation remains limited by small samples, heterogeneous manufacturing and isolation methods, inconsistent dose units, incomplete potency data, short follow-up, changing standards of care, and limited independent replication. By integrating clinical findings with chemistry, manufacturing, and controls, this chapter provides a product-specific framework for designing future MSC–EV trials and advancing the field from exploratory administration toward reproducible, mechanism-linked, and regulator-ready biologic development.

Keywords MSC–EVs · Extracellular vesicles · Exosomes · Clinical trials · Cell-free therapy · MSC secretome · Regenerative medicine · Safety · Tolerability · Potency assays · Dose standardization · Biodistribution · GMP · Regulatory translation · Therapeutic development

Clinical translation of MSC–EV-based therapy has progressed from isolated proof-of-concept use to a small but increasingly diverse body of human evidence. Published studies now include controlled investigations in chronic kidney disease, respiratory failure, pulmonary fibrosis, and diabetic foot ulcers, as well as small safety-oriented studies in stroke, complex regional pain syndrome, amyotrophic lateral sclerosis, and other conditions. The field is therefore no longer accurately described as purely preclinical. It is, however, still an early clinical development field in which most conclusions concern feasibility, short-term tolerability, route selection, and preliminary biological activity rather than confirmed efficacy.

N. M. Mussin, A. Tamadon, *Mesenchymal Stem Cells and Their Extracellular Vesicles*, https://doi.org/10.1007/978-3-032-36437-1_9

A central principle of this chapter is that an MSC–EV intervention must be evaluated as a defined product rather than as an interchangeable member of an "exosome" class. The therapeutic material administered in published studies has ranged from relatively enriched small EV preparations to mixed extracellular vesicle fractions and EV-enriched MSC secretomes. Products have been generated from MSCs derived from bone marrow, adipose tissue, umbilical cord, Wharton's jelly, placenta, and cord blood; isolated by ultracentrifugation, filtration, tangential flow filtration, chromatography, or proprietary workflows; formulated as frozen liquids, nebulized suspensions, infusions, topical gels, or locally delivered preparations; and dosed by particle number, protein mass, total volume, or formulation amount. These differences can alter identity, impurities, pharmacology, potency, and risk [23, 132, 222].

Terminology is equally important. In most clinical development settings, direct evidence of endosomal biogenesis is not available. The generic term extracellular vesicle, or an operational descriptor such as small extracellular vesicle, is therefore preferable to exosome unless biogenesis has been convincingly demonstrated. MISEV2018 and MISEV2023 emphasize transparent reporting of source material, separation and concentration procedures, EV-associated and non-EV markers, quantitative metrics, functional controls, and administration parameters [5, 71]. This nomenclature discipline is not semantic: it prevents apparent similarities in product names from obscuring clinically meaningful differences in composition and manufacturing.

The published clinical literature also requires an evidence hierarchy. A registered protocol demonstrates intent to investigate a product, not efficacy. An uncontrolled cohort can establish operational feasibility and identify common acute toxicities, but it cannot reliably separate treatment effects from disease course, supportive care, patient selection, or regression to the mean. Randomization improves causal inference, yet small randomized trials remain vulnerable to imprecision, multiplicity, and unstable subgroup findings. Confirmatory translation requires a reproducibly manufactured product, prospectively validated release and potency assays, adequate sample size, meaningful comparators, and replication across sites and patient populations.

Finally, the clinical program cannot be separated from chemistry, manufacturing, and controls (CMC). For complex biologics, the manufacturing process helps define the product. Changes in the producer cell bank, passage range, culture supplement, oxygen environment, harvest schedule, purification train, formulation, storage, or delivery device may alter critical quality attributes and biological activity. Clinical outcomes must therefore be interpreted together with batch-level product information, and later development should use comparability strategies rather than assuming that process changes preserve clinical identity [131, 159].

9.1 Current Clinical Landscape of MSC–EV Therapies

9.1.1 Published Human Studies and Registered Clinical Trials

The earliest frequently cited controlled human study evaluated cord blood MSC-derived EVs in 40 patients with stage III–IV chronic kidney disease. Twenty participants received 2 EV administrations and 20 formed a matched placebo group. The investigators reported improvements in estimated glomerular filtration rate, creatinine, blood urea, urinary albumin-to-creatinine ratio, and inflammatory mediators. Although the study was clinically important as an early demonstration of human administration, its small size, complex sequential delivery strategy, limited blinding information, and absence of later independent replication constrain causal and generalizable interpretation [162].

The COVID-19 pandemic accelerated respiratory development. A prospective nonrandomized cohort of 24 patients treated with a bone marrow MSC-derived EV-enriched product reported short-term feasibility and encouraging clinical signals, but lacked a concurrent randomized comparator [174]. A phase I study in 24 healthy volunteers subsequently evaluated a single nebulized dose of adipose MSC–EVs ranging from 2 × 10^8 to 16 × 10^8 particles and reported no serious adverse events through day 7, providing route-specific tolerability information rather than disease efficacy evidence [163]. A seven-patient phase IIa pilot used 2 × 10^8 nebulized particles daily for 5 days in severe COVID-19 and likewise reported acceptable immediate tolerability and exploratory imaging changes, but its uncontrolled design precluded efficacy inference [171].

A larger phase II multicenter, double-blind randomized trial enrolled 102 patients with COVID-19-associated moderate-to-severe respiratory failure and compared placebo with 10 mL and 15 mL doses of an EV-enriched bone marrow MSC secretome administered on days 1 and 4. No treatment-related adverse events were reported. The reduction in 60-day mortality in the overall intention-to-treat comparison for the 15 mL dose did not reach statistical significance; favorable findings were reported in post hoc age-defined analyses and therefore require prospective confirmation [161]. A later expanded-access analysis of 103 patients receiving up to three 15 mL infusions every 72 h reported one treatment-related grade 1 event and provided additional safety experience, but the absence of a control group prevents estimation of treatment efficacy [223].

More recent investigations have broadened the indication spectrum. A randomized, single-blind, placebo-controlled phase I study of nebulized umbilical cord MSC–EVs in 24 patients with pulmonary fibrosis reported no serious adverse events and exploratory improvements in pulmonary function and respiratory health status [165]. A randomized controlled study of a topical Wharton's jelly MSC-derived exosome gel enrolled 110 people with persistent diabetic foot ulcers and reported faster and more frequent healing in the treatment arm, although substantial attrition, per-protocol analyses, product-specific formulation effects, and the need for independent replication must be considered [166]. Small safety-oriented studies have also been published in malignant middle cerebral artery infarction, complex regional pain syndrome, and amyotrophic lateral sclerosis [224, 225, 226].

A 2024 systematic review and meta-analysis included 21 reports of human EV-based therapy across heterogeneous EV sources and indications. The pooled frequencies of serious adverse events and adverse events were low, but confidence intervals were wide, and the authors emphasized major heterogeneity in manufacturing, study design, characterization, and adverse event reporting. These pooled estimates should therefore not be interpreted as proof that all MSC–EV products are safe, particularly because the analysis combined different EV sources, routes, engineering states, and disease contexts [227]. A 2026 analysis of ClinicalTrials.gov records similarly documents expanding activity while underscoring heterogeneity and incomplete maturation of the field [228] (Table 9.1).

Table 9.1 Representative published human studies of MSC–EV-based or MSC–EV-enriched therapeutic products

Study (references)	Population and design	Product as reported	Route and dose	Principal finding and critical limitation
Nassar et al. (2016) [162]	40 patients with stage III–IV CKD; treatment and matched placebo groups	Cord blood MSC-derived EV fraction	Two administrations; IV followed by targeted renal arterial delivery; protein-based dosing	Renal and inflammatory marker improvements were reported. Small single-center study, complex delivery and no independent replication
Sengupta et al. (2020) [174]	24 patients with severe COVID-19; prospective nonrandomized cohort	Bone marrow MSC-derived EV-enriched secretome (ExoFlo)	Single 15 mL IV infusion	Feasibility and short-term clinical signals. No concurrent control and substantial pandemic-era confounding
Shi et al. (2021) [163]	24 healthy volunteers; phase I dose escalation safety study	Clinical-grade adipose MSC–EVs	Single nebulized dose, 2 × 10^8 to 16 × 10^8 particles	No serious adverse events through day 7. Safety in healthy volunteers does not establish disease efficacy or repeated dose safety
Dehghani et al. (2022) [225]	Five patients after malignant MCA infarction; open-label pilot randomized study	Placenta MSC-derived product described as exosomes	Local intraparenchymal injection during decompressive craniectomy	No serious local or systemic event reported. Very small sample and procedure-specific setting

(continued)

Table 9.1 (continued)

Study (references)	Population and design	Product as reported	Route and dose	Principal finding and critical limitation
Zhu et al. (2022) [171]	Seven patients with severe COVID-19; phase IIa single-arm pilot	Adipose MSC-derived product described as exosomes	Nebulized 2 × 10^8 particles daily for 5 days	Feasible and well tolerated; exploratory CT changes. Uncontrolled, late treatment and very small sample
Lightner et al. (2023) [161]	102 patients with COVID-19 respiratory failure; multicenter double-blind randomized trial	Bone marrow MSC-derived EV-enriched secretome (ExoFlo)	Placebo, 10 mL or 15 mL IV on days 1 and 4	No treatment-related adverse events. Overall mortality comparison was not significant; favorable post hoc subgroup signals need confirmation
Paicius et al. (2023) [226]	Ten patients with complex regional pain syndrome; open-label pilot	Bone marrow MSC-derived EV-enriched secretome (ExoFlo)	Two 15 mL IV infusions	No serious treatment-related adverse events and symptom improvements reported. Uncontrolled study with small sample
Zamanian et al. (2024) [164]	Critically ill patients with severe COVID-19; double-blind randomized controlled trial	Placental MSC-derived small EVs	IV, particle-normalized schedule over 2 days	Mortality and oxygenation signals were reported. Single-product, pandemic-context findings require replication and careful endpoint appraisal
Crose et al. (2024) [224]	Ten patients with ALS; 3-month safety-oriented study	Bone marrow MSC-derived EV-enriched secretome	Two 10 mL IV infusions 1 month apart	No serious treatment-related adverse events. Too small and brief to establish neurological efficacy or long-term safety
Li et al. (2025) [165]	24 patients with pulmonary fibrosis; randomized single-blind placebo-controlled phase I study	Umbilical cord MSC–EVs with multi-omic characterization	Nebulized administration	No serious adverse events; pulmonary function and questionnaire signals. Early phase, small sample and short follow-up

(continued)

Table 9.1 (continued)

Study (references)	Population and design	Product as reported	Route and dose	Principal finding and critical limitation
Kishta et al. (2025) [166]	110 patients with persistent diabetic foot ulcers; randomized three-arm controlled trial	Wharton's jelly MSC-derived exosome gel	Topical application with standard care weekly for 4 weeks	Higher healing rates and shorter healing time reported. Attrition, per-protocol analysis, and formulation-specific effects limit generalization
Sengupta et al. (2026) [223]	103 patients with severe/critical COVID-19 respiratory failure; multicenter expanded access	Bone marrow MSC-derived EV-enriched secretome (ExoFlo)	Up to three 15 mL IV infusions every 72 h	One related grade 1 event reported. Useful safety experience, but no control group and no causal efficacy inference

Note: Products described as "exosomes," "small EVs," EV fractions, or EV-enriched secretomes are not assumed to be interchangeable. Findings are summarized as reported by the original investigators and are interpreted at the individual product level

9.1.2 MSC–EV Product Types and Source Materials

Clinical products have been generated from MSCs derived from bone marrow, adipose tissue, umbilical cord, Wharton's jelly, placenta, and cord blood. Tissue source can influence producer cell expansion, immunobiology, surface molecules, RNA and protein cargo, and response to priming. Even products derived from the same tissue source may differ because of donor selection, passage range, culture density, oxygen tension, medium composition, serum or platelet lysate strategy, conditioning interval, and harvest frequency. The parental cell bank and its manufacturing history are therefore part of the medicinal product definition rather than background information.

The downstream process introduces a second layer of diversity. Differential ultracentrifugation may concentrate vesicles together with protein aggregates and other extracellular particles; filtration and tangential flow filtration can improve scale but may retain soluble macromolecules; chromatography can improve removal of selected impurities but changes recovery and product composition; and proprietary processes may yield an EV-enriched secretome rather than a highly purified EV fraction. A clinical report should state the operational product identity honestly and provide sufficient information to distinguish the administered material from conditioned medium, soluble secretome, a purified small EV fraction, and a combination product [5, 71].

Formulation and delivery further define the product. A frozen infusion formulation, nebulized suspension, topical gel, intra-arterial preparation, and surgically implanted local dose should not be expected to share pharmacokinetics or risk. The delivery device can also be product critical: aerosol droplet size, nebulizer efficiency, residual volume, shear, temperature, and collection losses affect the amount that reaches the respiratory tract. Similarly, a topical gel changes retention, hydration, diffusion, and wound surface exposure and may itself contribute to efficacy. Combination product development must therefore characterize the interaction between EVs and the vehicle or device.

Nomenclature should follow evidence. The word exosome should be reserved for preparations in which endosomal origin is supported; otherwise, extracellular vesicle or small extracellular vesicle is preferable. This distinction is especially important in clinical communication, because patients and clinicians may incorrectly assume that products carrying the same label are compositionally and therapeutically equivalent. MISEV-aligned operational naming, accompanied by full manufacturing and characterization information, is the more defensible approach [71].

9.1.3 Clinical Maturity and Evidence Hierarchy

The evidence base is unevenly distributed across the development continuum. Registry-only programs demonstrate scientific and commercial interest but contribute no clinical outcome data. Small feasibility studies can identify common immediate reactions, establish procedural practicality, and inform dose escalation. Controlled signal-finding studies can estimate effect direction and variance, but small samples and multiple outcomes make them vulnerable to unstable estimates. Confirmatory development requires prespecified primary endpoints, sufficient statistical power, multicenter execution, consistent background therapy, validated product release, and replication of the same product process configuration.

The majority of published MSC–EV studies remain in the safety/feasibility or early controlled signal-finding stages (Fig. 9.1). Randomized evidence exists, but it is concentrated in a limited number of indications and product platforms (Table 9.2). No finding should therefore be extrapolated to a generic MSC–EV class. A product manufactured from placental MSCs and dosed by particles cannot be assumed equivalent to an EV-enriched bone marrow secretome dosed by milliliters, even if both are administered intravenously for inflammatory lung disease.

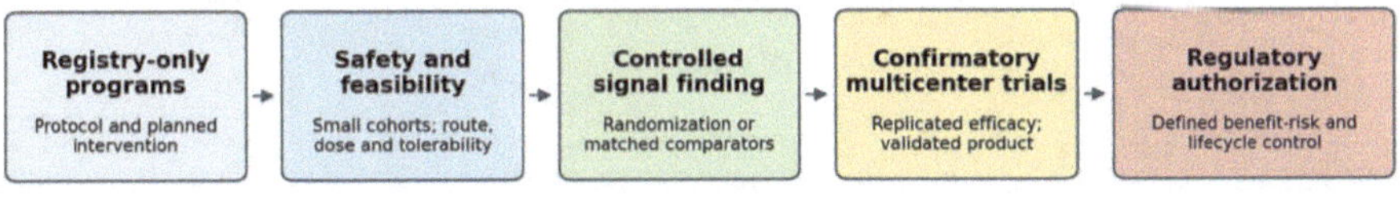

Fig. 9.1 Clinical maturity continuum for MSC–EV therapeutics. Most current human evidence remains within safety/feasibility and early controlled signal-finding stages. Progress should be assessed for a defined product, route, dose, and indication rather than for MSC–EVs as a presumed therapeutic class

Table 9.2 Selected registered or publication-linked MSC–EV clinical programs

Identifier	Indication/ population	Product and route	Evidence status through June 19, 2026	Interpretive point
NCT04313647	Healthy volunteers	Nebulized adipose MSC–EVs	Completed and published (Shi et al. 2021)	Defines single-dose inhalation tolerability, not disease efficacy
NCT04276987	Severe COVID-19	Nebulized adipose MSC-derived product described as exosomes	Completed and published (Zhu et al. 2022)	Small uncontrolled pilot; supports feasibility only
NCT04493242	COVID-19 respiratory failure/ARDS	IV bone marrow MSC–EV-enriched secretome	Completed and published randomized trial (Lightner et al. 2023)	Overall primary comparison and post hoc subgroup results must be separated
NCT04657458	Severe/ critical COVID-19 respiratory failure	IV bone marrow MSC–EV-enriched secretome	Expanded-access endpoint analysis published (Sengupta et al. 2026)	Adds exposure and safety experience without a concurrent comparator
ChiCTR2300075466	Pulmonary fibrosis	Nebulized umbilical cord MSC–EVs	Phase I randomized study published (Li et al. 2025)	Promising route-specific safety and signal finding; requires larger confirmation
NCT06812637	Persistent diabetic foot ulcer	Topical WJ–MSC-derived exosome gel plus standard care	Randomized controlled study published (Kishta et al. 2025)	Combination product and attrition effects should be addressed in replication
NCT04213248	Chronic GVHD-associated dry eye	Umbilical cord MSC-derived exosome eye drops	Registry-based program; publication linkage should be rechecked	Illustrates local ocular development and the need for route-specific endpoints
NCT05176366	Medically refractory ulcerative colitis	IV EV-enriched bone marrow MSC secretome	Registry record; status and results should be rechecked before final production	Highlights expansion into chronic inflammatory disease without established efficacy

Note: Registry records are dynamic. Recruitment status, sponsor information, endpoints, and publication linkage must be checked immediately before the final book is submitted for production

Evidence appraisal should also account for sponsorship, analytic flexibility, and publication linkage. Several reports involve proprietary products and sponsor-affiliated authors. This does not invalidate the data, but it increases the importance of protocol transparency, independent replication, complete adverse event reporting, and publication of neutral or negative outcomes. Trial registries should be linked unambiguously to publications, and changes in recruitment status, endpoints, or manufacturing should be documented.

A recent systematic review and meta-analysis by Mussin et al. [229] synthesized the available human evidence concerning MSC-derived extracellular vesicle therapies. The analysis reinforced that the emerging clinical signals remain product, route, and indication specific and highlighted substantial heterogeneity in parental cell source, EV preparation, product characterization, dose units, administration schedules, comparators, and outcome definitions. Accordingly, the available evidence supports continued evaluation of defined MSC–EV products in controlled clinical trials but does not yet justify class-wide conclusions regarding efficacy or safety.

9.2 Therapeutic Indications Studied in Humans

9.2.1 Respiratory Diseases, ARDS, and COVID-19-Related Indications

Respiratory disease currently provides the densest human MSC–EV evidence, largely because the COVID-19 emergency created both urgent clinical need and accelerated access pathways. The biological rationale was strong: MSC-derived products can modulate inflammatory signaling, macrophage behavior, endothelial and epithelial barrier function, and tissue repair. The clinical record, however, spans very different products and routes and must be interpreted against rapidly changing standards of care, viral variants, vaccination status, intensive care practices, and enrollment thresholds.

Nebulization is attractive because it attempts to place the product near the airway and alveolar compartment while reducing systemic exposure. The healthy-volunteer MEXVT study provided an important first step by demonstrating that single doses up to 16 × 10^8 adipose MSC–EV particles were tolerated over short follow-up [163]. The subsequent MEXCOVID pilot treated seven patients with 2 × 10^8 particles daily for 5 days and reported no prespecified inhalation-associated events or immediate instability. Imaging improvement in several patients was hypothesis generating, but treatment occurred late in the disease course, and there was no control group [171]. These studies support the operational feasibility of nebulization, not a general conclusion that inhaled MSC–EVs improve ARDS outcomes.

Intravenous development has used both EV fractions and EV-enriched secretome products. The 2020 cohort reported no acute safety signal and encouraging survival in 24 patients, but its uncontrolled design and early-pandemic setting make historical comparisons unreliable [174]. The 102-patient EXIT COVID-19 randomized trial substantially improved evidentiary quality. It demonstrated acceptable short-term tolerability for two infusions and a numerical survival difference for the 15 mL arm, but the primary intention-to-treat mortality comparison was not statistically significant. The more favorable age-defined analyses were post hoc and should be treated as a basis for prospective stratification rather than confirmation [161].

A double-blind randomized study of placental MSC-derived small EVs reported favorable mortality and oxygenation findings in severe COVID-19 [164]. Another randomized study compared perinatal MSCs and their EVs in COVID-19-associated ARDS, illustrating the difficulty of attributing efficacy when cell and EV platforms, background treatment, and multiple arms are evaluated within one trial [230]. These reports are valuable, but independent replication with harmonized products and endpoints is needed before clinical recommendations can be made.

The 2026 expanded-access analysis adds experience with repeated 15 mL infusions in 103 patients and reported 1 treatment-related grade 1 event. Because expanded access prioritizes treatment and safety rather than comparative efficacy, the mortality and hospital course observations cannot establish benefit [223]. This distinction is essential: large uncontrolled cohorts may strengthen detection of common acute toxicity but remain weak evidence for treatment effectiveness.

Future respiratory trials should move beyond broad labels such as COVID-19 pneumonia or ARDS and use biologically and clinically coherent phenotypes. Baseline severity, time from symptom onset, inflammatory state, ventilation mode, organ failure, corticosteroid and immunomodulator exposure, and pulmonary fibrosis burden may all modify response. Primary outcomes should be prospectively prioritized, and subgroup analyses should be limited, biologically justified, and powered. For inhaled products, delivered lung dose and device performance should be integrated into the clinical protocol rather than treated as ancillary technical details.

9.2.2 Pulmonary Fibrosis and Chronic Inflammatory Conditions

Pulmonary fibrosis represents a strategically important shift from acute rescue to chronic disease modification. In the 2025 phase I investigation, 24 participants were randomized in a single-blind, placebo-controlled design to evaluate nebulized umbilical cord MSC–EVs. The product was supported by extensive preclinical and multi-omic characterization, and no serious adverse events were reported. Signals in forced vital capacity, maximal voluntary ventilation, and respiratory health status were encouraging [165]. Nevertheless, the sample was small, follow-up was limited, and phase I studies are not designed to establish durable antifibrotic efficacy.

Fibrotic disease trials should distinguish transient anti-inflammatory or bronchodilator-like changes from modification of extracellular matrix remodeling. Clinically meaningful assessment requires serial pulmonary function, exercise capacity, oxygen requirement, exacerbations, hospitalization, high-resolution computed tomography, validated patient-reported outcomes, and sufficiently long follow-up to estimate slope of decline. Biomarkers may support mechanism, but they should not substitute for functional and clinical outcomes unless validated as surrogates.

Other chronic inflammatory indications, including inflammatory bowel disease and ocular surface disease associated with chronic graft-versus-host disease, appear in trial registries. These programs are scientifically plausible, yet registry presence alone does not show efficacy. Each indication will require a potency assay that reflects the intended biology. An immune suppressive assay suitable for inflammatory bowel disease, for example, may not predict epithelial repair in dry eye disease. The field should resist using a single generic potency test across unrelated indications.

9.2.3 Kidney Disease and Systemic Inflammatory or Neurological Disorders

The chronic kidney disease study by Nassar et al. [162] remains one of the earliest controlled examples of MSC–EV administration in humans. The reported improvements in renal function and inflammatory mediators provided a strong translational signal, but the study also demonstrates why modern reporting standards are necessary. Sequential intravenous and targeted renal arterial administration makes route attribution difficult; the EV preparation included multiple vesicle classes; and the study predates MISEV2023 and current expectations for batch-level potency and impurity characterization. A confirmatory kidney program should use a reproducible route, standardized background nephrology care, prespecified renal endpoints, and longer follow-up [162].

Small studies in complex regional pain syndrome and amyotrophic lateral sclerosis have evaluated repeated intravenous administration of an EV-enriched bone marrow MSC secretome. Both reported acceptable short-term tolerability and exploratory clinical changes, but neither was designed or powered to establish efficacy [224, 226]. These reports are best viewed as feasibility studies that inform later trial design. Pain outcomes are especially vulnerable to expectancy, regression to the mean, concomitant treatment, and rater effects; neurological diseases require sufficient duration to distinguish true modification of disease trajectory from measurement variability.

The five-patient stroke study evaluated local intraparenchymal administration during decompressive craniectomy. No serious injection site reaction, excessive edema, or hematoma was reported, supporting procedural feasibility in a highly selected surgical setting [225]. The intervention cannot be generalized to routine ischemic stroke care because administration was invasive, the sample was extremely small, and the clinical course was dominated by the underlying malignant infarction and surgery. Future neuroregenerative trials should separate procedural risk from product risk and incorporate imaging, functional outcomes, neuroinflammation markers, and long-term neurological surveillance.

9.2.4 Wound Healing, Diabetic Foot Ulcers, and Topical Applications

Topical delivery is attractive because it can maximize local exposure while limiting systemic distribution. It also creates a combination product problem: the vehicle or biomaterial can affect moisture, retention, diffusion, protection from proteolysis, and wound cell behavior. A well-designed topical trial therefore requires a vehicle-matched placebo and standardized background wound care, debridement, off-loading, infection management, vascular assessment, and glycemic control.

The 2025 diabetic foot ulcer study randomized 110 participants to standard care plus WJ–MSC-derived exosome gel, standard care alone, or standard care plus a vehicle formulation. The investigators reported a higher complete healing proportion and shorter healing time in the treatment group [166]. The inclusion of both standard care and vehicle controls is a strength. Important limitations include substantial attrition, differences between randomized and analyzed populations, reliance on per-protocol analysis, a single-center product and preparation process, and the possibility that the formulation or intensive wound management contributed to outcomes. Replication should use intention-to-treat analysis, centralized blinded wound assessment, standardized digital planimetry, and prespecified handling of amputation, infection, revascularization, and missing data.

Future topical MSC–EV studies should prioritize complete epithelialization confirmed at more than one visit, time to durable closure, recurrence, infection, amputation, pain, health-related quality of life, and resource use. Product dose should include both EV metrics and formulation concentration per wound area. Local safety should include irritation, allergic reaction, infection, delayed healing, abnormal granulation, and systemic exposure where repeated or extensive application is planned.

9.2.5 *Emerging Indications and Early Exploratory Trials*

The wider EV clinical landscape includes ocular, dermatological, inflammatory, neurological, pulmonary, and engineered EV programs. A case report described placental MSC-derived exosome-enriched EVs for chronic cutaneous graft-versus-host disease, illustrating potential local regenerative and immunomodulatory effects but providing no comparative evidence [231]. Registry programs have also explored dry eye disease, ulcerative colitis, osteoarthritis, and other conditions.

Indication expansion should be governed by product mechanism fit rather than by the broad pleiotropy attributed to MSCs. Before entering a new disease, developers should show that the clinical product—not merely a research preparation from the same cell source—has relevant potency in disease-linked assays, reaches the intended compartment, and has a route-specific safety margin. A platform may support multiple indications, but the potency assay, dose, formulation, patient selection, and endpoints may need to change. Each such change creates a comparability question.

Registry analyses are useful for mapping activity but are vulnerable to duplicate records, outdated status, missing results, inconsistent terminology, and weak linkage between registered interventions and publications. Therefore, a final book chapter should separate published human outcomes from registered-only programs and timestamp all registry conclusions [228].

Regenerative endodontics represents an emerging local application of MSC–EV therapy. Jafari et al. [232] described the application of human umbilical cord MSC-derived exosomes in a pulpectomized human tooth, providing an early clinical feasibility signal for dental pulp–dentin regeneration. However, because the evidence is derived from a single case, it should be considered hypothesis generating rather than confirmatory. Controlled studies are required using standardized EV characterization, clearly defined treatment procedures, appropriate comparators, and objective radiographic, vascular, sensory, and tooth survival endpoints.

9.3 Safety, Tolerability, and Clinical Monitoring

9.3.1 *Infusion, Inhalation, and Local Administration Risks*

MSC–EVs are nonreplicating and avoid several hazards intrinsic to viable cells, including uncontrolled cellular proliferation, ectopic cell differentiation, and obstruction by large infused cells. This does not make them intrinsically safe. Their proteins, lipids, nucleic acids, surface phospholipids, residual process components, co-isolated particles, and formulation excipients can produce biological effects. Safety assessment must therefore be product and route specific.

For intravenous administration, immediate monitoring should include infusion reactions, fever, hypersensitivity, hemodynamic instability, oxygenation changes, complement activation, coagulation abnormalities, thromboembolic events, and organ function changes. Systemically administered EVs commonly encounter the mononuclear phagocyte system, and in vivo identity may be modified by protein corona formation, which can alter cellular uptake and biodistribution [158]. Preclinical biodistribution and toxicology should use the intended clinical formulation and route, not a laboratory surrogate.

For inhaled or nebulized products, safety monitoring should include cough, wheeze, bronchospasm, airway irritation, desaturation, respiratory deterioration, and device-related delivery variability. The nominal dose placed in a nebulizer is not the same as the dose emitted, inhaled, deposited, or retained. Device performance, aerosol characteristics, breathing pattern, disease severity, airway obstruction, and ventilatory support can all influence exposure. Repeated dose programs should assess anti-product responses and cumulative pulmonary effects.

For topical, intra-articular, intra-arterial, intraparenchymal, or other local routes, procedural and product risks must be separated. Local administration may reduce systemic exposure but can produce tissue injury, infection, pressure effects, inflammation, abnormal repair, or off-target escape. When an EV product is combined with a gel, scaffold, patch, or delivery device, both components and their interaction require assessment.

9.3.2 Immunogenicity, Coagulation, and Procoagulant Concerns

Allogeneic MSC–EV products are often described as low immunogenicity, but repeated human dosing data remain limited. EV preparations can display major histocompatibility complex proteins, immune regulatory ligands, donor-associated molecules, and adsorbed plasma proteins. Clinical programs should evaluate whether repeated dosing induces anti-donor or anti-product immune responses when biologically plausible, particularly if producer cells are inflammatory primed or genetically modified.

Hemocompatibility is a specific concern for intravascular products. Phosphatidylserine exposure and tissue factor activity can promote coagulation, and direct experimental work has demonstrated procoagulant activity in some umbilical cord MSC–EV preparations [170]. The magnitude of risk is product dependent and can be influenced by cell source, culture state, isolation method, particle concentration, and route. Developers should not infer non-thrombogenicity from the absence of cells.

A risk-based clinical package may include tissue factor antigen and activity, thrombin generation assays, clotting time, platelet interaction, complement activation, and clinical biomarkers selected for the product and population. Patients with severe inflammation, malignancy, immobility, vascular disease, or concurrent prothrombotic treatment may require additional precautions. Hemocompatibility should be integrated into release and comparability strategies when intravenous use is intended [23].

9.3.3 Tumor-Related Concerns and Long-Term Follow-Up

MSC–EVs cannot form a cellular graft or replicate, but they can influence angiogenesis, cell survival, immune surveillance, extracellular matrix remodeling, and the tumor microenvironment. Depending on producer cell state and cargo, MSC-derived signals may theoretically support or suppress tumor-associated processes. This bidirectionality argues against a generic claim that cell-free therapy is free of tumor risk [167].

Tumor-related assessment should be proportionate to exposure, indication, route, dosing frequency, patient age, and product biology. For short local wound treatment, long-term systemic surveillance may be limited; for repeated systemic immunomodulation in chronic disease, longer follow-up may be warranted. Clinical protocols should define how incident malignancy, recurrence, unexplained mass lesions, persistent immunosuppression, and delayed organ toxicity will be captured.

Long-term follow-up is also needed for durability and delayed immune effects. Most published studies have follow-up measured in days or months, which is insufficient to characterize rare events or chronic repeated dosing. A safety database should combine trial follow-up, pharmacovigilance, registry linkage where appropriate, and systematic batch traceability.

9.3.4 Donor Screening, Sterility, Endotoxin, and Clinical Release Testing

Clinical-grade MSC–EV products require a qualified and traceable producer cell system, controlled manufacturing, and predefined release criteria. At minimum, clinical reports should describe the following:

- Donor eligibility, tissue procurement, infectious disease screening, consent, and traceability.
- MSC identity, purity, passage range, growth characteristics, senescence or genetic stability controls as relevant, and master/working cell bank strategy.
- Culture medium, supplements, EV depletion method, oxygen tension, priming, harvest window, and collection frequency.
- Clarification, concentration, purification, formulation, fill–finish, storage, thawing or reconstitution, and delivery device handling.
- Particle concentration and size distribution, total protein or other complementary quantitative metrics, morphology, EV-associated markers, and non-EV/host cell impurity markers.
- Sterility, mycoplasma, endotoxin, adventitious agent controls, residual reagents, and particulate or aggregate assessment.

- A quantitative, mechanism-linked potency assay or justified assay matrix with predefined acceptance criteria.
- Lot number, release date, stability status, deviations, and comparability status for every administered batch.

Rohde et al. [132] and Gimona et al. [131] provide practical foundations for converting EV characterization into clinical manufacturing and potency strategies. The exact numerical limits cannot be universal; they must be justified for the specific product and clinical use. Safety alerts and enforcement actions involving unapproved products marketed as exosomes further demonstrate why research-grade characterization is not an adequate substitute for regulated manufacturing and clinical oversight [148, 233, 234].

9.4 Efficacy Signals and Current Evidence Gaps

9.4.1 Early Clinical Signals Across Organ Systems

Across published studies, early signals have been reported for respiratory recovery and mortality, pulmonary function and health status, kidney function measures, pain and patient-reported symptoms, neurological safety, and diabetic wound closure. These findings are clinically encouraging because they show that MSC-derived acellular products can be manufactured, delivered, and monitored in humans. They should nevertheless be described as product-specific signals rather than proof of a common mechanism or class-wide benefit.

The strongest signal is not necessarily the largest numerical effect. Evidentiary weight depends on randomization, allocation concealment, blinding, completeness of follow-up, prespecification, endpoint validity, statistical precision, and replication. A modest effect in a well-controlled, reproducibly manufactured program may be more informative than a dramatic change in a small uncontrolled cohort. Likewise, a biologically plausible biomarker change does not guarantee meaningful patient benefit.

The EV field should explicitly distinguish exploratory, supportive, and confirmatory evidence. Exploratory findings generate hypotheses; supportive findings strengthen mechanism or consistency; confirmatory findings require adequate power and control of multiplicity. This vocabulary would reduce the tendency to overstate small early-phase studies.

9.4.2 Endpoint Heterogeneity and Lack of Standardized Comparators

Clinical endpoints have varied from mortality and ventilation-free days to oxygenation, computed tomography scores, renal biomarkers, pulmonary function, wound closure, disability scales, pain scores, and quality of life. Such variation is expected across indications, but even within one indication, studies often use different definitions, time windows, and analysis populations. Cross-trial synthesis is therefore difficult.

Endpoints should follow the intended mechanism and disease stage. In acute respiratory failure, mortality may be clinically definitive but statistically inefficient in small trials; ordinal recovery, ventilation-free days, organ support-free days, and time to discharge may be informative if prespecified and analyzed appropriately. In pulmonary fibrosis, serial functional decline and exacerbation may be more meaningful than a short-term imaging or biomarker change. In wound healing, durable complete closure is stronger than percentage area reduction alone.

Comparator design is equally important. Acute trials require standardized background care and documentation of time-dependent treatments. Topical trials should include a vehicle control when the formulation may influence healing. Local surgical trials need a comparator that separates procedure from product. When placebo is not feasible, blinded outcome adjudication and objective endpoints become more important.

9.4.3 Dose Metrics, Potency Correlation, and Exposure–Response Uncertainty

Published MSC–EV trials have reported doses as particles per kilogram, total particle number, protein mass, product volume, or formulation amount. These units are not interchangeable. Particle counting instruments detect non-EV particles and differ in sensitivity; protein mass can be dominated by co-isolated material; product volume provides little information without concentration and potency; and cell-equivalent dosing depends on upstream yield and culture conditions.

At least two complementary quantitative metrics should be reported when feasible, together with the measurement method, dilution, calibration, particle-to-protein ratio, total volume, route, schedule, and number of administrations. For inhaled products, nominal, emitted, and estimated delivered dose should be distinguished. For topical products, dose per unit wound area and formulation concentration should be reported. For systemic products, body size normalization may be considered but should not replace direct product metrics.

The long-term solution is potency-anchored dosing. A defined functional assay can connect administered quantity to biological activity and provide a basis for lot comparison. Potency should be mechanism linked, quantitative, stability indicating, and feasible for release. A single assay may be insufficient for pleiotropic products, but an indiscriminate battery of assays is also unsuitable. The assay matrix should be limited to orthogonal tests that jointly control the clinically relevant activity [131].

Exposure–response relationships remain largely unknown. Systemic EVs commonly accumulate in the liver and spleen, while local and inhaled routes aim to increase compartment-specific exposure. Protein corona formation and disease state changes can alter cellular uptake and distribution [158, 173]. Future trials should integrate nonclinical pharmacokinetics, biodistribution, pharmacodynamic biomarkers, and clinical response rather than escalating dose solely by convention.

9.4.4 Limitations of Small Trials and Nonrandomized Evidence

Most human MSC–EV reports remain small, single-center, uncontrolled, or exploratory. These designs can be appropriate for first-in-human safety and feasibility, but they are highly susceptible to selection bias, confounding, changing supportive care, incomplete outcome ascertainment, and publication bias. Rare toxicities cannot be excluded by the absence of events in cohorts of 5, 7, 10, or 24 participants.

Small randomized trials improve internal validity but often generate imprecise estimates and may be dominated by chance imbalance. Post hoc subgroup findings are particularly fragile and should be labeled explicitly. Attrition can also change the meaning of a randomized comparison when analyses exclude substantial numbers of participants. Intention to treat should generally be primary, with per-protocol analyses used as supportive sensitivity analyses.

The 2024 meta-analysis of 21 EV therapy reports found low pooled adverse event frequencies, but heterogeneity and wide confidence intervals limit product-specific inference [227]. Pooling cannot correct for inconsistent product definitions, dose metrics, routes, or reporting. The next phase of evidence generation should prioritize fewer, better-defined, multicenter programs over continued proliferation of small, noncomparable studies.

9.5 Recommendations for Future MSC–EV Clinical Trial Design

9.5.1 MISEV-Aligned Clinical Reporting

Clinical reporting should extend MISEV principles from experimental rigor to investigational product transparency. The manuscript and trial report should define the producer cell source and bank, culture and conditioning conditions, separation and purification process, product identity and impurity profile, dose metrics, formulation, storage history, route, device, release specifications, and potency result for the administered lots [5, 71].

The product name should be operationally accurate. Exosome should not be used as a marketing synonym for a small particle preparation. If a study administers an EV-enriched secretome, it should be described as such. If soluble mediators remain in the product, they should be measured or acknowledged. Clinical authors should avoid implying that a positive result validates all MSC–EVs.

Reports should also provide batch numbers or an anonymized lot map, manufacturing site, whether one or multiple lots were administered, deviations, and whether process changes occurred during recruitment. Negative and neutral outcomes should be published. Trial registry identifiers, protocol versions, statistical analysis plans, and data-sharing statements should be provided.

9.5.2 Integration of CMC, Potency Assays, and Clinical Endpoints

CMC and clinical development should proceed as one program. The intended clinical mechanism should guide selection of critical quality attributes, potency assays, patient population, pharmacodynamic biomarkers, and clinical endpoints. An immunomodulatory intravenous product may require an immune cell or cytokine response assay and hemocompatibility testing. A wound healing product may require epithelial migration, angiogenesis, or macrophage modulation assays together with formulation release testing. A pulmonary fibrosis product may require assays reflecting epithelial protection, inflammatory modulation, or antifibrotic activity.

Clinical lots should meet predefined potency criteria, and exploratory analyses should examine whether batch-level potency or other critical attributes correlate with exposure and outcome. Such analyses can identify clinically meaningful CQAs and improve future specifications. They must be planned carefully because small trials are underpowered for complex batch response modeling.

Manufacturing changes should trigger a documented comparability assessment. Analytical similarity may be sufficient for low-risk changes, but changes that affect source material, culture biology, purification, formulation, or potency may require nonclinical or clinical bridging. The practical goal is to preserve the identity of the product tested, not merely to maintain a similar particle count.

9.5.3 Route-Specific Trial Design

Route is a first-order design variable. Intravenous trials should emphasize systemic exposure, hemocompatibility, infusion reactions, clearance organ safety, and thromboembolic surveillance. Inhaled trials should qualify the nebulizer, characterize aerosol output, define emitted and delivered dose, standardize breathing or ventilatory conditions, and monitor bronchospasm and oxygenation. Topical trials should control wound care and vehicle effects. Local invasive trials should distinguish procedural complications from product toxicity.

Dose escalation rules should be based on nonclinical safety, potency, expected exposure, and route-specific constraints. Repeated dosing requires justification of interval, accumulation, immune response, and stability after preparation. A maximum feasible dose is not automatically a biologically optimal dose. Adaptive designs may be useful when dose metrics and early pharmacodynamic responses are uncertain, provided decision rules are prespecified.

A route-specific development plan should also define the relevant pharmacodynamic compartment. Blood biomarkers may be suitable for systemic immune modulation but may poorly reflect local lung or wound activity. Bronchoalveolar, imaging, tissue, tear film, synovial fluid, or wound fluid measures may be more informative when collection is feasible and ethically justified.

9.5.4 Priorities for Multicenter Randomized Controlled Trials

The next generation of multicenter trials should include the following design priorities (Table 9.3):

1. A fully defined and reproducibly manufactured clinical product with a locked or prospectively controlled process.
2. Centralized or harmonized lot release testing and a quantitative mechanism-linked potency assay.
3. Prospectively justified dose units, route, schedule, and delivery device.
4. A biologically coherent and clinically relevant patient population with major prognostic factors captured at baseline.
5. Randomization, concealment, blinding where feasible, and an appropriate placebo, vehicle, or standard care comparator.
6. A clinically meaningful primary endpoint with a prespecified hierarchy for secondary, biomarker, and patient-reported outcomes.
7. Sample size justification based on realistic effect estimates and allowance for attrition and multiplicity.
8. Independent safety monitoring, route-specific stopping rules, and prespecified adjudication of serious events.
9. Intention-to-treat analysis, transparent handling of missing data, and restrained subgroup analyses.
10. Batch-level traceability, comparability documentation, and analysis of clinically relevant product attributes.
11. Follow-up sufficient for durability, repeated dose safety, immunogenicity, coagulation, organ toxicity, and tumor-related concerns.
12. Commitment to registration, protocol and statistical analysis plan availability, and publication regardless of outcome.

Table 9.3 Minimum design and reporting dataset for future MSC–EV clinical trials

Domain	Minimum requirement	Preferred advanced approach
Source and parental cells	Tissue source, donor eligibility, passage, identity, culture and harvest conditions	Qualified master/working cell bank, genetic stability, and source-specific functional characterization
Manufacturing	Complete concentration, purification, formulation, storage and fill–finish description	Closed GMP-compatible process with validated in-process controls and lifecycle comparability
Identity and purity	Particle size/count, morphology, EV-associated markers, and selected impurity markers	Orthogonal bulk and single-particle analytics, calibrated counting, and quantitative impurity profile
Dose	At least one validated quantitative unit plus volume, route, schedule, and measurement method	Particle number, protein mass, ratio metrics, potency-normalized dose, and delivered dose estimate
Potency	Biological activity demonstrated with justified assay	Validated quantitative mechanism-linked release assay correlated with stability and clinical response
Route/device	Full administration procedure and device description	Device qualification, emitted/delivered dose validation, and route-specific biodistribution package
Safety	Sterility, mycoplasma, endotoxin, and prespecified adverse event monitoring	Hemocompatibility, complement, immunogenicity, biodistribution, organ toxicity, and long-term surveillance
Comparator	Standard care or placebo where feasible	Vehicle-matched control, standardized background therapy, and blinded central adjudication
Endpoints	Prespecified safety and clinically relevant efficacy endpoints	Hierarchical clinical, functional, patient-reported, and mechanism-linked pharmacodynamic endpoints
Statistics	Prospective sample size and analysis plan	Multicenter design, multiplicity control, missing data strategy, sensitivity analyses, and constrained subgroup testing
Traceability	Lot identification and release status	Batch-level CQA/potency linkage to exposure and outcome
Reporting	Registry ID, protocol, manufacturing summary, and complete adverse event reporting	MISEV-aligned product dossier, statistical analysis plan, data sharing, and publication independent of outcome

Abbreviations: *CQA* critical quality attribute, *EV* extracellular vesicle, *GMP* good manufacturing practice, *MISEV* minimal information for studies of extracellular vesicles

These priorities align with modern Good Clinical Practice, risk-based quality management, and the principle that clinical evidence must remain linked to the product actually administered [235]. Figure 9.2 summarizes the integrated development framework.

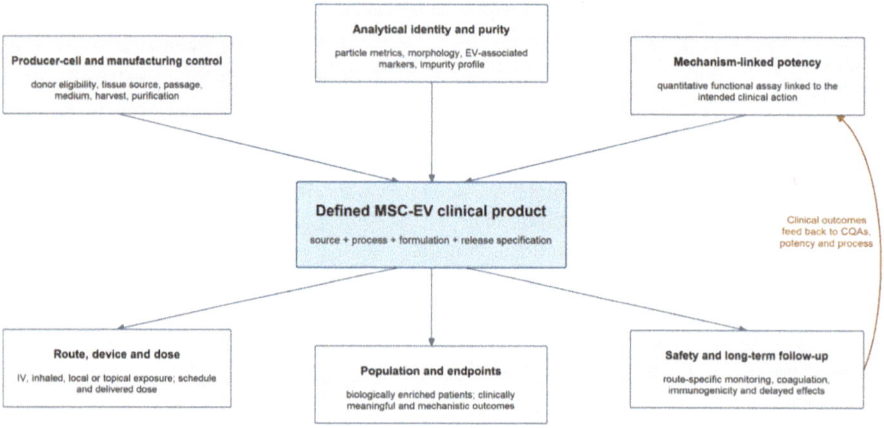

Fig. 9.2 Integrated development framework for MSC–EV clinical trials. Product definition, potency, exposure, patient selection, endpoints, and safety are connected, with clinical outcomes feeding back to critical quality attributes and process control

Chapter 10
Manufacturing, GMP, Regulatory, and Commercialization Pathways

Abstract This chapter addresses manufacturing, regulatory, and commercialization pathways for MSC–EV-based therapeutics. It discusses GMP-compliant production, process scalability, and comparability challenges. Regulatory frameworks from major agencies are analyzed, with emphasis on potency assays, quality control, and lifecycle management. The chapter also explores industrial translation and economic considerations for bringing EV-based products to market.

Keywords GMP · Manufacturing · Regulatory pathways · Commercialization · Quality control

Manufacturing readiness is a regulatory and commercial strategy, not a late-stage technical task: the strongest sources converge on a lifecycle view in which product understanding → process control → analytical control → managed change are the backbone of GMP compliance, credible filings, and scalable supply. This is explicit in the continuous manufacturing and lifecycle management frameworks (ICH Q12/Q13) and in the modern analytics lifecycle (ICH Q14/Q2(R2)), which collectively reward early definition of what is "established," what is "changeable," and what evidence is needed when either the process or the assay evolves [236].

For complex biologics and advanced therapies, recent FDA draft and final guidance emphasize that potency cannot be "fixed" by a single late assay. Instead, regulators increasingly expect a potency assurance strategy grounded in quality risk management: potency is protected through process design and controls, material controls, in-process testing, and fit-for-purpose potency release assays with meaningful acceptance criteria, progressively strengthened as knowledge, clinical data, and manufacturing experience accumulate [25].

Scale-up and site/network expansion are repeatedly identified as high-risk inflection points for regulatory objections, especially when comparability arguments are weak or assays are not sufficiently suitable to detect meaningful differences. FDA's 2023 draft guidance on manufacturing changes/comparability for cellular and gene therapy products and EMA's 2019 Q&A on ATMP comparability provide convergent expectations: risk-assess the change, select relevant attributes and methods

N. M. Mussin, A. Tamadon, *Mesenchymal Stem Cells and Their Extracellular Vesicles*, https://doi.org/10.1007/978-3-032-36437-1_10

(often beyond routine release), define acceptance criteria before execution, and design studies sufficiently robust to support a definitive conclusion [26].

Sterile/aseptic controls remain a frequent "make-or-break" constraint for industrialization. The revised EU GMP Annex 1 formalizes contamination control strategy (CCS), expands expectations for barrier technology and monitoring, and sets phased implementation dates (with full applicability by August 2024 for specified provisions). Even when a product is not terminally sterilized, Annex 1 principles (risk management, facility design, qualification, monitoring, personnel controls) shape expectations for low-bioburden intermediates and sterile steps used across modalities [237].

Commercialization pathways are increasingly constrained by manufacturability economics (labor, cleanroom footprint, throughput, QC cycle time) and by post-approval change burden. Recent CAR-T manufacturing literature and economic analyses show automation can reduce cost and footprint, but only when paired with analytics, data integrity, and standardized control strategies that survive scaling and transfer. In parallel, recent regulatory "lessons learned" analyses of cell/gene therapy approvals highlight recurring quality objections around comparability, potency assay validation, specifications, and stability—precisely the topics this chapter treats as a single connected system rather than separate workstreams [238].

10.1 GMP-Compliant Production

GMP-compliant production is best understood as a closed-loop system that links (i) what the product must be (quality attributes), (ii) how the process reliably produces it (process controls), (iii) how results are proven and recorded (data integrity + documentation), and (iv) how changes are managed without eroding safety, quality, or supply. This lifecycle framing is embedded in ICH Q12 (post-approval lifecycle management), in ICH Q9(R1) (quality risk management), and, for advanced therapies, in FDA's potency assurance and comparability guidance that explicitly connect process design/control, material controls, and in-process testing with final lot release [236] (Fig. 10.1).

A contemporary GMP chapter must therefore treat "GMP" as more than cleanrooms and SOPs. For sterile and aseptic operations, the revised EU GMP Annex 1 elevates the contamination control strategy (CCS) into a required, actively reviewed,

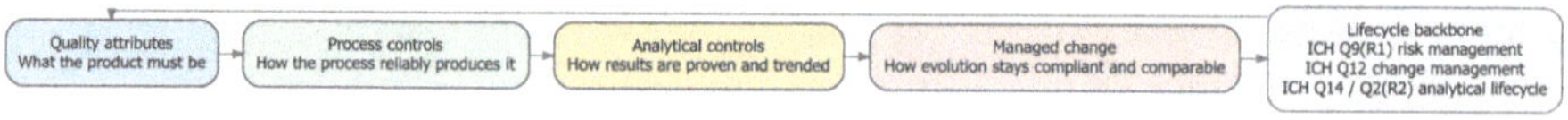

Fig. 10.1 GMP production as a closed-loop control architecture. GMP-compliant production is shown as a closed-loop control system linking product quality attributes, process controls, analytical controls, and managed change. The figure emphasizes that manufacturing readiness is not a late technical activity but a lifecycle strategy in which process understanding, data integrity, and change governance jointly support compliance, filings, and scalable supply

continually improving strategy that integrates facility design, utilities, environmental monitoring, personnel practices, and process choices, with defined implementation milestones. This reframes compliance from "meeting discrete clauses" to demonstrating a coherent, risk-based control architecture [237].

Advanced therapy manufacturing amplifies classic GMP challenges because starting materials may be variable, batch sizes may be small, logistics may be time critical, and manufacturing may be distributed across sites. The EU's GMP guidelines specific to ATMPs explicitly adapt GMP expectations to ATMP characteristics, including risk-based application of GMP principles and heightened attention to traceability, materials, and process understanding. In parallel, FDA's gene therapy CMC guidance emphasizes providing sufficient CMC information in the IND to assure safety, identity, quality, purity, and strength (including potency), with expectations growing as development progresses [18].

10.1.1 *Key Challenges*

A first recurring challenge is contamination and sterility assurance (Fig. 10.2): Annex 1 expands the scope and depth of expectations around barrier technology, cleanroom classification, environmental and process monitoring, and CCS

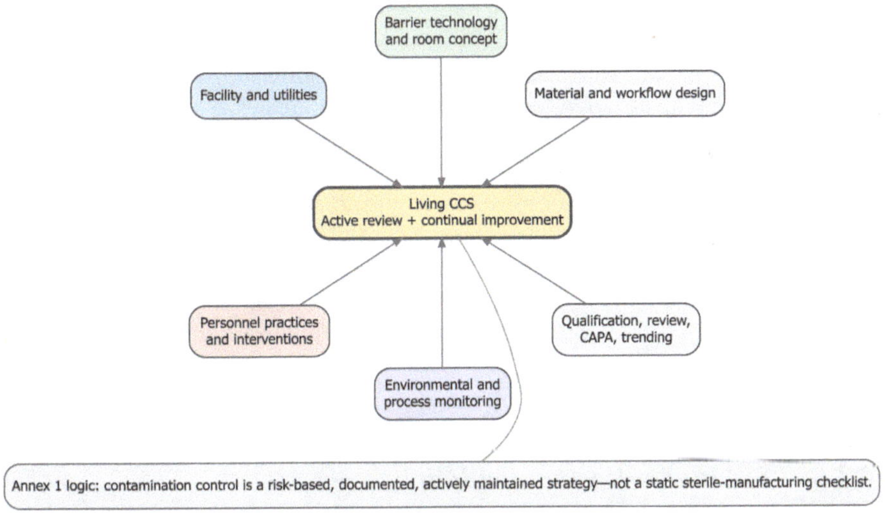

Fig. 10.2 Contamination control strategy for aseptic and sterile manufacturing. This figure organizes the contamination control strategy as an actively maintained system integrating facility and utility design, personnel practices, barrier technology, environmental and process monitoring, material flow, qualification, and CAPA review. It reflects the Annex 1 expectation that contamination control is a living, risk-based governance framework rather than a collection of disconnected sterile manufacturing requirements

governance—raising the compliance bar for both new and legacy facilities and for organizations relying on manual interventions [237].

A second challenge is data integrity as a prerequisite for regulatory trust. FDA's data integrity Q&A clarifies that CGMP compliance requires reliable and accurate data, with risk-based controls appropriate to system complexity and intended use. Recent real-world events underscore why regulators react strongly when integrity of manufacturing- or quality-related datasets is questioned [239].

A third challenge is viral safety and adventitious agent control for biologics derived from characterized cell lines. The updated ICH Q5A(R2), adopted by FDA as guidance, refreshes risk-based principles and mitigation strategies for viral safety in light of scientific and technology progress, including updated expectations for data submitted in marketing applications. Even when a program is not a classical monoclonal antibody, the revision is relevant wherever cell substrates, helper viruses, or viral vector–derived components intersect with purification and clearance strategies [240].

10.1.2 Best Practices

A best-practice GMP operating model in 2016–2026 guidance is risk based by design, not reactive. ICH Q9(R1) explicitly addresses known failure modes of quality risk management—subjectivity, unclear "formality," and weak links between risk outputs and decisions—making it particularly relevant for organizations scaling rapidly or operating multi-site networks. Using Q9(R1) as the organizing logic for CCS (EU Annex 1), potency assurance (FDA draft guidance), and manufacturing change risk assessment (FDA comparability draft) creates a common language across QA, manufacturing science, and regulatory strategy [241].

A second best practice is explicit definition of what is "established" and how it may change. ICH Q12 provides a harmonized framework for post-approval CMC changes, anchored in the concepts of established conditions and proactive change management tools, intended to improve predictability and enable continual improvement without destabilizing supply. Treating ICH Q12 as a "design input" early in development is increasingly a differentiator at commercialization, because it reduces future regulatory friction around process and analytical evolution [236].

10.1.3 Actionable Recommendations

Operationalize GMP as a "control loop" with measurable performance. Implement a living CCS (where sterile steps exist) that is tied to formal QRM, periodic management review, and continual improvement triggers [237]. Treat data integrity as a process capability: define audit trail review expectations, system access governance, and risk-based review frequency aligned with FDA's data integrity guidance; include integrity checks as release gating criteria for batch record review, not only as audit findings [239].

For biologics, build viral safety strategy early using Q5A(R2) concepts: justify cell substrate qualification approaches, detection/monitoring methods, and clearance claims in a way that maps to specific unit operations and reporting expectations for submissions [240].

10.2 Batch Consistency and Scale-Up

Batch consistency is the practical test of whether "process understanding" is real. In modern guidance, consistency is demonstrated through validated control strategies, defined ranges and criteria, and—critically—comparability evidence when manufacturing changes occur. FDA's 2023 draft comparability guidance for cellular and gene therapy products frames change management as a lifecycle requirement: risk assessment should be done for all manufacturing changes, and comparability studies should be performed when the change could adversely affect product quality. It also acknowledges a hard truth: for complex products, analytical studies may not be sufficient alone, and additional nonclinical or clinical evidence may sometimes be needed [26].

From a scale-up perspective, the same guidance emphasizes study design elements that directly affect credibility: comparability requires careful selection of relevant attributes, methods, acceptance criteria, and statistics; where feasible, studies should use full-scale lots, and if full scale is infeasible, a data-driven risk assessment and CPP/CQA justification are expected. Importantly, comparability may require measuring attributes not used for routine release—an explicit signal that release testing alone is rarely enough to support major scale or site changes [26].

EMA's 2019 Q&A on ATMP comparability provides a complementary European lens: comparability is needed following manufacturing changes or when introducing additional manufacturing sites, and late-stage changes should trigger scientific advice interactions due to potential impact on the acceptability of clinical data. Together with FDA expectations, this converges on a principle that should be treated as doctrine: "manufacturing changes are clinical changes unless proven otherwise" [242].

A modern scale-up chapter should treat continuous manufacturing (CM) as a strategic option rather than a niche method. ICH Q13 provides harmonized scientific and regulatory considerations for developing, operating, and lifecycle-managing CM for drug substances and drug products, including for converting batch processes to CM—explicitly framing CM as compatible with new products, generics, biosimilars, and therapeutic proteins (with applicability principles extendable to other biologic entities). This matters commercially because CM can change throughput, control paradigms, and batch definition—thus requiring intentional design of control strategy and release logic [243].

10.2.1 Key Challenges

The most acute scale-up challenge for cellular therapies is that "batch" may be patient specific (autologous) and capacity is constrained by labor, facility grade, QC cycle time, and logistics. Recent CAR-T manufacturing reviews describe high labor requirements and bottlenecks spanning manufacturing technology, QC/QA, and global capacity—constraints that directly translate into commercial lead time and cost [238] (Fig. 10.3).

For viral vectors, scale-up challenges include upstream yields, downstream purification capacity, and maintenance of product quality attributes across scale and process changes. Recent AAV manufacturing reviews emphasize the centrality of rigorous process characterization, validated scale-down models, and DoE-informed operating ranges as prerequisites for robust control strategies that survive translation [244].

A cross-cutting challenge in both batch and CM paradigms is that definition and control of "the batch" must be aligned to release testing and traceability. FDA's recent guidance on complying with 21 CFR 211.110 explicitly notes that continuous manufacturing still requires defining batches to be tested for release and to satisfy CGMP requirements, reinforcing that CM does not relax release accountability—it shifts how evidence is assembled [245].

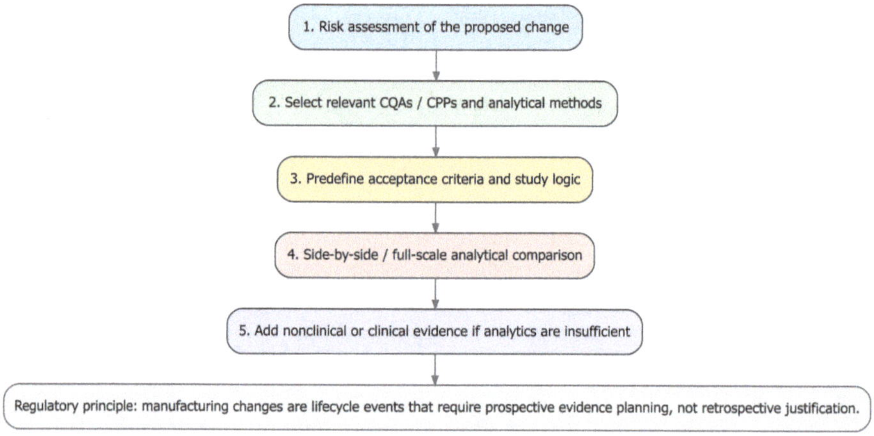

Fig. 10.3 Comparability evidence ladder for scale-up, site transfer, and process change. The figure presents comparability as a graded evidence ladder beginning with formal risk assessment, followed by selection of relevant CQAs and CPPs, predefined study logic, side-by-side analytical comparison, and escalation to nonclinical or clinical evidence when analytics alone are insufficient. It captures the modern regulatory principle that manufacturing changes are lifecycle events requiring prospective evidence planning rather than retrospective justification

10.2.2 Best Practices

Use ICH Q12 to pre-design scale-up for future change: specify established conditions and propose disciplined change management that uses prior knowledge and risk-based categorizations. This improves submission quality and reduces "late surprises" when commercial realities force facility or material changes [236].

For multi-site manufacturing (especially for cell therapies), follow the logic in FDA's CAR-T guidance: multi-step procedures and multiple biological materials create lot variability risk; therefore, robust process controls, in-process and lot release testing, and intentional comparability management are crucial—particularly because changes during development are common [246].

10.2.3 Actionable Recommendations

Define a "comparability readiness package" before scale-up begins: a living document that lists expected future changes (materials, equipment, site expansion), the risk rationale for each, candidate comparability attributes/methods, and predefined acceptance criteria/statistics where feasible (updated as knowledge grows) [26]. Adopt side-by-side testing as a default for major transitions when possible, consistent with FDA's recommendation to make comparability data readily interpretable by controlling assay and site variables [26]. When selecting CM as a scale strategy, map Q13 concepts early to batch definition, material traceability, transient disturbance management, and how in-process controls will replace or complement end product testing [243].

10.3 Potency Assays and Release Criteria

Potency sits at the intersection of science (mechanism and biological activity), GMP (control and release), and regulation (evidence of clinical effectiveness consistency). FDA's December 2023 draft guidance defines a potency assurance strategy as a multifaceted approach that reduces risks to potency through manufacturing process design and control, material control, in-process testing, and potency lot release assays, with the goal that every released lot has the capacity to achieve the intended therapeutic effect [243] (Table 10.1 and Fig. 10.4).

A critical modern reframing in this draft guidance is that potency release assays are not required to "perfectly mimic" mechanism of action; rather, assay design should quantify potency-related CQAs that are at risk, and at least one bioassay is recommended when a potency-related CQA tied to biological activity can only be measured effectively biologically. This shifts assay strategy from "single

Table 10.1 Comparing potency assay types

Potency assay type (typical use cases)	Strengths (pros)	Limitations (cons)	Validation/control expectations
Cell-based functional bioassay	Closest direct measure of biological activity; can cover complex functional attributes when surrogates are insufficient [25]	Often higher variability and longer cycle time; sensitive to cell reagents, operator effects, and drift; QC throughput can constrain release [25]	Use Q14 to design for robustness and control strategy; validate fit-for-purpose performance characteristics per Q2(R2); use system suitability and reference materials; acceptance criteria must mitigate potency risk and should not be "report only" [25]
Reporter gene assay (cell based)	Quantitative, scalable, and often more reproducible than complex primary functional assays; supports moderately high throughput [247]	Still sensitive to cell line stability and transfection/transduction variability; may measure a proximal pathway signal rather than full MoA [25]	Define ATP/intended purpose (Q14); validate selectivity/specificity, accuracy, precision, range, and robustness (Q2(R2)); maintain reagent and cell bank controls; manage changes with risk-based equivalence/bridging [247]
Binding assay	High precision and throughput; good for identity/relative binding strength proxies; useful as part of assay matrix for multi-CQA potency assurance [25]	May be a surrogate not directly linked to biological effect; risk if binding does not correlate with clinical activity [25]	Q2(R2) validation of accuracy/precision/selectivity/range; Q14-driven development data can support validation. Require reference standards and system suitability; control reagent lots and method transfers [248]
Enzymatic activity assay (where relevant)	Often quantitative and relatively fast; can serve as strong potency surrogate when enzyme activity is direct mechanism [248]	Not applicable to all products; may ignore delivery/function steps for complex modalities [25]	Validate per Q2(R2) with purpose-driven criteria; define and control critical assay parameters (Q14) and ensure meaningful acceptance criteria tied to potency assurance risks [248]
Transduction/infectivity assay (viral vectors, functional infectious titer)	Measures functional delivery capability; more biologically informative than genome copy number alone [249]	Can be time-consuming and variable; may depend on cell substrate and assay design; may constrain release timelines [249]	Risk-based inclusion in potency assurance; Q14 robustness exploration; Q2(R2) validation; strong controls for cells/reagents; change control and comparability planning since assay drift can confound conclusions [25]

(continued)

Table 10.1 (continued)

Potency assay type (typical use cases)	Strengths (pros)	Limitations (cons)	Validation/control expectations
qPCR/ddPCR genome titer (viral vector genomes)	High throughput; quantitative; supports consistency monitoring and can be part of potency matrix [248]	Measures genome copies, not biological activity; may miss critical functional failures [25]	Validate selectivity, accuracy/precision, and range (Q2(R2)); Q14 development data informs controls (primer/probe design, inhibitors); pair with at least one bioassay when biological activity CQAs can't be sufficiently controlled otherwise [250]
Flow cytometry phenotypic surrogates	Rapid, informative for identity/viability/phenotype; supports in-process control and release panels [246]	Often surrogate; variability from gating strategy/instrument settings; inter-site harmonization difficult in multi-site networks [246]	Treat as analytical procedures with lifecycle controls: method standardization, system suitability, qualification/validation as fit for purpose, and equivalence testing for method transfer [250]

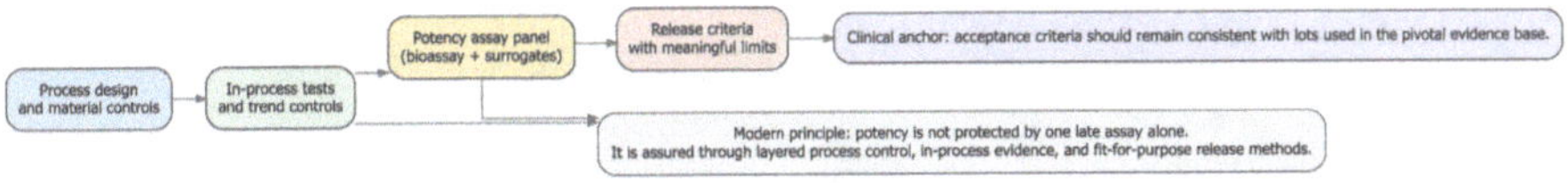

Fig. 10.4 Potency assurance matrix linking process, assays, and release decisions. Potency assurance is depicted as a layered matrix spanning process design and material controls, in-process testing, potency assay panels, and scientifically justified release criteria. The figure reflects the current regulatory expectation that potency cannot be protected by a single late assay alone, but instead must be assured through an integrated control strategy anchored to clinically proven product performance

all-encompassing potency number" toward assay panels/matrices justified by risk and CQA coverage, supported by a control strategy that mitigates residual risk elements [25].

Release criteria, in this modern view, are not merely pass/fail specifications; they are risk mitigations that must be scientifically defensible. FDA's draft potency assurance guidance explicitly discourages "report/for information only" acceptance criteria for release assays because they do not add to potency assurance, and it argues for quantitative lower limits (and, when needed, upper limits) informed by manufacturing data, nonclinical studies, and/or clinical experience. Notably, it also links licensed-product acceptance criteria to consistency with lots used in the pivotal evidence base, tying release directly to the clinical "anchor" [25].

Assay development and validation expectations have also been modernized and harmonized across ICH regions. ICH Q14 describes science- and risk-based approaches for developing and maintaining analytical procedures fit for purpose, while ICH Q2(R2) provides validation elements as part of the analytical procedure lifecycle, explicitly linking development knowledge (Q14) to validation strategy and enabling abbreviated validation when scientifically justified. FDA adopted both as 2024 guidance to facilitate science-based regulatory evaluation and risk-based post-approval change management of analytical procedures [247].

10.3.1 Key Challenges

A recurring regulatory challenge is that potency assays may be inherently variable, and establishing meaningful release acceptance criteria can be delayed when MoA is still evolving. FDA's comparability draft highlights that potency assay development difficulty can delay setting release acceptance criteria, yet excluding potency from comparability can be problematic—creating a "catch-up" risk late in development [26].

Operationally, potency assays often become the rate-limiting step for product release and therefore for commercial scalability, particularly in autologous cell therapy where turnaround time is clinically and logistically constrained. Recent CAR-T manufacturing literature emphasizes QC/QA and release testing as major capacity bottlenecks in real-world delivery systems [238].

10.3.2 Best Practices

Use FDA's potency assurance framing to design a layered defense: if certain potency-related CQAs can be robustly controlled by process and material controls, potency release testing can focus on the remaining high-risk CQAs; but at least one bioassay is generally recommended where biological activity cannot be adequately assured by non-biological measures alone [25]. Integrate ICH Q14 and Q2(R2) explicitly into potency plans: define intended purpose/ATP, use development knowledge to justify validation strategy, and build post-approval method change flexibility by documenting controls and risk rationales that translate into regulatory confidence [247].

10.3.3 Actionable Recommendations

Define potency-related CQAs early (even if provisional) and re-run formal risk assessment at each development stage to decide which CQAs require lot release assays versus process/in-process controls, aligning with FDA's guidance for progressive strengthening of potency assurance [25]. Make acceptance criteria clinically anchored by design: plan analytic and clinical bridging studies that explicitly link release assay readouts to lots in the pivotal evidence base, consistent with

FDA's recommendation that licensed-product acceptance criteria ensure consistency with clinically proven lots [25].

Build assay transfer as a comparability activity: for multi-site or outsourced testing, pre-specify equivalence/bridging approaches for assay transfer so that site expansion does not invalidate historical comparability datasets [26].

10.4 Regulatory Considerations and Product Classification Pathways

Regulatory strategy is an integrated plan for what the product is, how it is controlled, and how evidence is assembled across the lifecycle—and classification determines which pathway, dossier expectations, and inspection regimes apply (Table 10.2). For many advanced therapies in the FDA framework, the gene therapy CMC guidance and CAR-T guidance both situate these products within the existing biologics framework (IND → BLA) and position CMC as the evidence package that assures product safety and quality during clinical development, with increasing submission depth as trials expand [249].

In the EU/EMA context, ATMP-specific guidance similarly emphasizes structured data packages for clinical trial applications and marketing authorizations, recognizing gene therapy medicinal products, somatic cell therapy medicinal products, tissue-engineered products, and combined ATMPs under the ATMP legal umbrella. The EMA gene therapy guideline focuses on marketing authorization requirements (quality, safety, efficacy) and notes that revisions reflect issues arising from marketing authorizations and scientific advice—an implicit signal that common deficiencies are now "known pitfalls" [251].

In Japan [249], PMDA guidance for ensuring quality and safety of gene therapy products lays out technical expectations for quality and safety assurance for in vivo gene therapy products and ex vivo genetically modified human cell therapies among regenerative medical products, reinforcing that gene therapy–specific CMC considerations are treated as a distinct technical domain. PMDA also publishes position documents on enabling technologies that shape regulatory expectations for modernized production paradigms [249].

A unifying modern regulatory theme is that risk-based lifecycle management is the "common backbone" across modalities and regions. ICH Q12 creates the structure for predictable post-approval CMC change management; ICH Q13 provides a harmonized approach for continuous manufacturing lifecycle; ICH Q14/Q2(R2) provides an analytical lifecycle approach; and ICH Q9(R1) refines how risk methods should be used to reduce subjectivity and protect product availability. Together, these allow regulators to grant flexibility—but only when sponsors demonstrate that knowledge, control, and change governance are mature [236].

Table 10.2 Comparing major regulatory guidance documents

Region/body	Guidance document (year)	Scope (what it governs)	Relevance to this chapter
FDA	Potency Assurance for Cellular and Gene Therapy Products (Draft, 2023)	Science- and risk-based potency assurance strategy; potency assays; acceptance criteria; assay lifecycle control [25]	Primary source for potency strategy, assay selection, acceptance criteria, and lifecycle maturation [25]
FDA	Manufacturing Changes and Comparability for Human Cellular and Gene Therapy Products (Draft, 2023)	Lifecycle approach to managing/reporting manufacturing changes; comparability study design; risk assessment; analytical methods and statistics [26]	Core source for batch consistency, change management, scale-up comparability evidence [26]
FDA	Considerations for the Development of CAR-T Cell Products (Final, 2024)	CAR-T-specific CMC recommendations, including process control, analytical testing, and comparability; also nonclinical/clinical guidance [246]	High-specificity template for cell therapy manufacturing control and multi-site considerations [246]
FDA	CMC Information for Human Gene Therapy INDs (Final, 2020)	CMC content expectations for gene therapy INDs to assure safety, identity, quality, purity, and strength (including potency) [249]	Foundational for gene therapy manufacturing packages and early-phase GMP strategy [249]
FDA	Q2(R2) Validation of Analytical Procedures (Final, 2024; ICH)	Analytical procedure validation principles across lifecycle; links to Q14; supports flexible change management when justified [250]	Defines validation expectations and the evidence basis for release and comparability assays [250]
FDA	Q14 Analytical Procedure Development (Final, 2024; ICH)	Science–/risk-based analytical development and lifecycle maintenance; foundations for method change management [252]	Key for designing robust potency and release methods and enabling defensible evolution [252]
FDA	Data Integrity and Compliance With Drug CGMP (Final, 2018)	Data integrity expectations under CGMP; Q&A format; risk-based controls and governance [239]	Establishes regulatory trust prerequisites for manufacturing and QC data used in submissions [239]
EMA/EU (EC GMP)	EU GMP Annex 1: Manufacture of Sterile Medicinal Products (Revised, 2022; applicability milestones through 2024)	Sterility assurance expectations; CCS; facility, barrier tech, monitoring, personnel, process controls [237]	Governs aseptic/sterile operations relevant to fill–finish and sterile intermediates across modalities [237]
EU (European Commission)	GMP specific to ATMPs (2017)	Risk-adapted GMP framework for ATMP manufacturing under EU law [18]	Key for industrial translation of ATMP platforms in Europe (facility, materials, traceability) [18]

(continued)

Table 10.2 (continued)

Region/body	Guidance document (year)	Scope (what it governs)	Relevance to this chapter
EMA	Gene therapy medicinal products guideline (2018)	Quality, nonclinical, and clinical aspects for gene therapy MAAs; updates reflect clinical and regulatory experience [251]	Aligns CMC development and product characterization with EU expectations for GTMPs [251]
EMA	Investigational ATMPs in clinical trials guideline (2025)	Data requirements/structure for ATMP clinical trial applications [32]	Drives CMC expectations pre-commercialization in the EU (clinical stage) [32]
EMA	Q&A on comparability for ATMPs (2019)	ATMP comparability expectations after manufacturing changes or new sites [242]	Practical comparability lens; complements FDA comparability principles for CMC strategy [242]
EMA	Process validation for finished products (Revision 1, 2016)	Process validation information/data expectations in dossiers; lifecycle framing [130]	Anchors validation expectations for traditional products and informs biologics case by case [130]
ICH	Q12 Lifecycle management (Step 4, 2019)	Framework for post-approval CMC change management; established conditions; lifecycle tools/enablers [236]	Central to commercialization and "change without re-approval crisis" [236]
ICH	Q13 Continuous Manufacturing (Step 4, 2022)	Scientific/regulatory considerations for CM development, operation, and lifecycle management [243]	Key for scale strategy and modern manufacturing paradigms affecting release and control [243]
ICH	Q14 Analytical procedure development (Step 4, 2023)	Science–/risk-based analytical development and maintenance [247]	Directly informs potency assay development as a lifecycle capability [247]
ICH	Q2(R2) Analytical validation (Step 4, 2023)	Validation elements integrated into lifecycle; links to Q14 [248]	Validation basis for release criteria, comparability assays, and stability-indicating methods [248]
ICH	Q9(R1) Quality Risk Management (Step 4, 2025; FDA adoption 2023)	Refines QRM to reduce subjectivity, clarify formality, and address supply risks [241]	Backbone for CCS, potency assurance, and comparability risk assessments [253]
ICH/FDA	Q5A(R2) Viral safety evaluation (Step 4, 2023; FDA guidance 2024)	Viral safety risk-based principles; testing, clearance, and submission expectations [240]	Critical for biologics viral safety packages and some vector-derived products [240]

(continued)

Table 10.2 (continued)

Region/body	Guidance document (year)	Scope (what it governs)	Relevance to this chapter
PMDA	Ensuring the Quality and Safety of Gene Therapy Products (2019)	Technical requirements for quality and safety for gene therapy products (in vivo and ex vivo genetically modified cells) [254]	Japan-specific CMC expectations for regenerative/gene therapy products [254]
PMDA	Views on Applying Continuous Manufacturing (2018)	PMDA perspective on CM concepts, considerations, and emerging expectations [255]	Useful for global CM strategy and Japan alignment in industrial translation [255]

10.4.1 Key Challenges

A central regulatory challenge across regions is misalignment between product evolution and evidence continuity (Fig. 10.5). New analyses of cell and gene therapy approvals report frequent quality objections around manufacturing comparability, potency assay validation, specifications, and stability—suggesting that many programs still treat these as compliance checkboxes rather than as "core evidence" [256].

10.4.2 Best Practices

Use harmonized ICH frameworks as the "spine" of a global dossier: Q12 for change management, Q13 for CM concepts, Q14/Q2(R2) for analytical lifecycle, and Q9(R1) for risk governance. This reduces region-specific rework and helps maintain a single quality narrative across submissions [236].

10.4.3 Actionable Recommendations

Treat classification as a manufacturing strategy input: define early whether the product will be regulated as a biologic/ATMP/regenerative product, because that determines not only the dossier format but the inspection expectations, comparability burden, and potency assurance strategy maturity required by pivotal stage and commercial filings [249].

Build a "regulatory evidence map" that explicitly links every proposed major process or assay change to (i) risk assessment rationale, (ii) planned comparability attributes/methods, and (iii) reporting pathway aligned with lifecycle stage—anticipating that late changes can undermine clinical interpretability if comparability is weak [26].

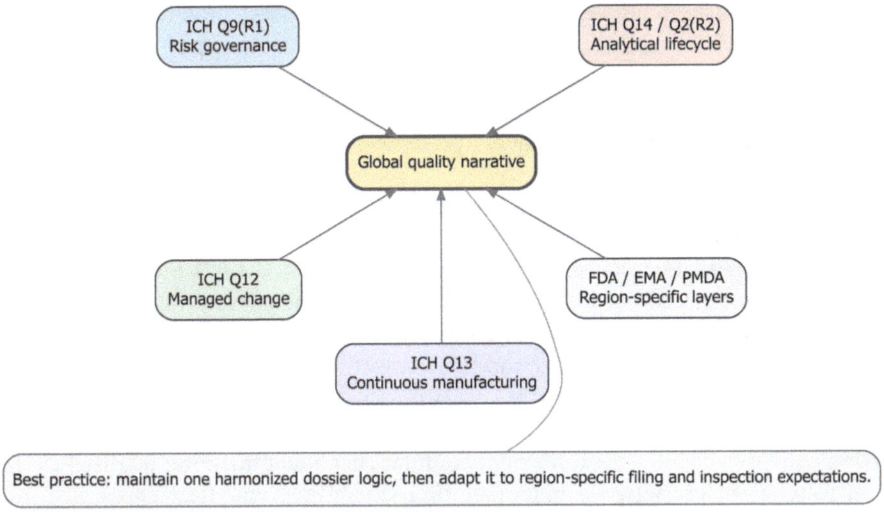

Fig. 10.5 Global regulatory spine for manufacturing, analytics, and lifecycle management. This figure maps the harmonized regulatory backbone for industrial translation, showing how ICH Q9(R1), Q12, Q13, and Q14/Q2(R2) support a single global quality narrative, while regional frameworks from FDA, EMA, and PMDA add modality- and jurisdiction-specific layers. It emphasizes that robust global strategy depends on maintaining one integrated quality logic across risk management, analytical lifecycle, manufacturing evolution, and post-approval change planning

10.5 Commercialization and Industrial Translation

Commercialization is the process of converting a clinical manufacturing system into a repeatable industrial value chain with predictable cost, capacity, quality, and regulatory flexibility. For advanced therapies, technology translation is often limited not by scientific feasibility but by industrial feasibility: facility constraints, labor intensity, QC cycle time, batch failure risk, and supply chain fragility. Recent CAR-T manufacturing reviews describe global capacity limits and the difficulty of translating novel platforms into GMP production at scale, while economic analyses show that automation can reduce costs and facility footprint—yet requires upfront technology investment and robust in-process/QC analytics to remain compliant and performant [238].

Commercial success is also increasingly shaped by how well manufacturing change is managed post-approval. ICH Q12's explicit goal is to facilitate predictable, efficient management of post-approval CMC changes, promoting innovation and continual improvement while strengthening quality assurance and reliable supply. The practical implication for industrial strategy is that manufacturing design should anticipate future scale, site, and method evolution and embed those expectations into the initial control strategy and submission posture where possible [236].

A major commercialization risk is that deficiencies in manufacturing governance can surface publicly and disrupt launch trajectories. FDA's data integrity guidance makes clear that reliable data is foundational to CGMP compliance, and FDA's public statement on data accuracy issues following a landmark gene therapy approval illustrates how integrity failures can create post-approval scrutiny and operational disruption. These events translate directly into commercialization risk: delayed scale expansion, intensified oversight, reputational damage, and loss of payer/provider confidence [239].

Case studies of 2017–2023 approvals show that "industrial translation" increasingly requires combining (i) sophisticated CMC and analytics and (ii) distribution/network readiness. For example, FDA approvals of early CAR-T therapies in 2017 established commercial precedent for individualized manufacturing and risk controls, while later approvals of gene therapies for hemophilia and sickle cell disease reflect maturation of gene delivery and genome editing modalities. Meanwhile, the first full FDA approval of an mRNA COVID-19 vaccine in 2021 demonstrates an alternative industrial translation archetype: rapid scale, extensive global manufacturing networks, and the need for high confidence in manufacturing quality [257].

The relationship between manufacturing capability, analytical capability, and regulatory lifecycle control is summarized in Fig. 10.6.

10.5.1 Key Challenges

Industrial translation commonly fails at the handoff between "process works" and "process scales with evidence." The most frequent quality objections reported for cell and gene therapies—comparability, potency assay validation, specifications, and stability—map directly to areas where organizations often underinvest until late stages, when fixes are expensive and may destabilize clinical interpretability [256].

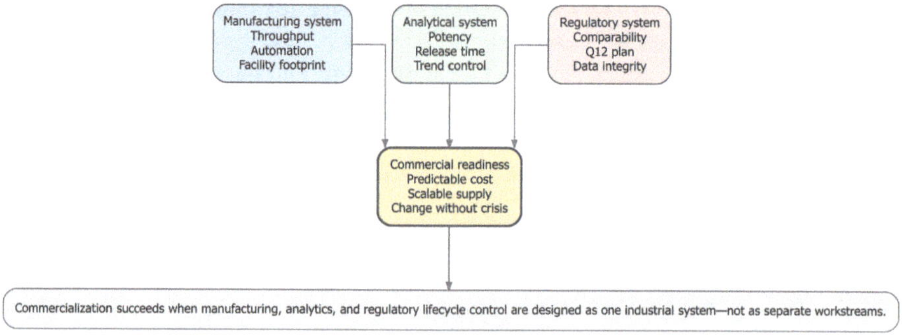

Fig. 10.6 Industrial translation as a three-system integration problem. Commercial readiness is shown as the outcome of integrating three interdependent systems: manufacturing capability, analytical capability, and regulatory lifecycle control. The figure highlights that predictable cost, scalable supply, and post-approval flexibility emerge only when throughput, potency and release analytics, comparability planning, and data integrity are designed as one industrial system rather than as separate workstreams. Representative approval and industrial translation examples illustrating manufacturing, CMC, quality, and lifecycle management lessons are summarized in Table 10.3

Table 10.3 Case studies of recent approvals or industrial translations

Product (approval example)	Reference region and pathway	Why it matters for manufacturing/CMC	Key lessons for industrial translation
Kymriah (tisagenlecleucel)	FDA approval (2017, biologics pathway) [257]	Early commercial CAR-T highlighted individualized lot control and the release/QC/ logistics interdependence [238]	Design manufacturing and QC as a single operational "clock"; anticipate capacity bottlenecks in release testing and chain of identity [238]
Yescarta (axicabtagene ciloleucel)	FDA approval (2017, biologics pathway) [258]	Reinforced the need for lot-to-lot consistency controls in multi-step, variable starting material processes [246]	Scale strategy must include comparability plans for inevitable process evolution; multi-site manufacturing requires harmonized assays and governance [246]
Luxturna (voretigene neparvovec-rzyl)	FDA approval (2017, gene therapy product page and supporting documents) [259]	In vivo AAV gene therapy commercialization highlighted the challenge of producing high-quality viral vectors under GMP [244]	Build vector manufacturing capability early; ensure potency/ functional assays and comparability strategies can handle scale and site changes [26]
Zolgensma (onasemnogene abeparvovec-xioi)	FDA press announcement (2019) [260]	Demonstrates scale and scrutiny: high-impact pediatric gene therapy; post-approval data integrity concerns became public [261]	Commercialization risk includes data governance: Integrity failures can trigger regulator/public actions; embed integrity controls and independent verification [239]
Hemgenix (etranacogene dezaparvovec-drlb)	FDA approval documentation hub (2022) [262]	Illustrates mature gene therapy licensing and the importance of lot release compliance language in approval documentation [263]	For commercial gene therapies, ensure release testing strategy and acceptance criteria are fully defensible; plan for long-term CMC change needs (Q12) [25]
Roctavian (valoctocogene roxaparvovec)	EMA conditional marketing authorization (2022) and FDA approval (2023) [264]	Shows multi-region strategy and different authorization types; highlights need for aligning CMC evidence across regions [251]	Global industrialization requires harmonized comparability and analytic lifecycle controls to avoid region-specific rework and delays [242]
Casgevy + Lyfgenia	FDA press announcement (2023, first cell-based gene therapies for sickle cell; genome editing + lentiviral) [265]	Marks maturation of ex vivo genome editing and viral gene addition; reinforces potency/ comparability/acceptance criteria significance for complex modalities [25]	Potency must be assured by layered strategy; comparability planning is crucial because assays and processes evolve as MoA and manufacturing knowledge grow [25]

(continued)

Table 10.3 (continued)

Product (approval example)	Reference region and pathway	Why it matters for manufacturing/CMC	Key lessons for industrial translation
Comirnaty (COVID-19 vaccine, mRNA)	FDA full approval (2021, biologics license) [266]	Demonstrates industrial translation at global scale and the centrality of manufacturing quality to public confidence [266]	Scale demands robust analytic lifecycle (Q14/Q2(R2)), change management discipline (Q12), and data integrity as an operational capability—Not a compliance afterthought [252]

A second challenge is that capacity constraints are often analytics driven, not bioreactor driven: release cycle times, assay variability, and batch record review can dominate lead time. This is particularly acute for autologous therapies and for high-variability functional assays, where automation in manufacturing must be matched by maturity in analytical and QA systems to realize its economic benefits [238].

10.5.2 Best Practices

Treat commercialization as a "three-system integration": (1) manufacturing technology (throughput, automation, facility), (2) analytical lifecycle (Q14/Q2(R2) with potency assurance strategy), and (3) regulatory lifecycle (Q12/Q9(R1) for managed change and risk governance). Programs that integrate these early reduce the probability that scale-up forces uncontrolled process or assay changes that then trigger comparability crises [236].

10.5.3 Actionable Recommendations

Design a commercialization "CMC exit criteria" checklist that must be met before pivotal trials: validated or late-stage qualified potency assays with meaningful acceptance criteria, defined comparability protocol templates, data integrity controls demonstrably effective, and a Q12-aligned post-approval change plan (especially if automation/site expansion is expected) [25]. When selecting make/buy (in-house vs CDMO) strategies, require that partners can support the analytical lifecycle (method transfer equivalence, reference materials governance, system suitability, trending) and change management discipline; otherwise, outsourcing can increase comparability risk rather than reduce operational burden [26]. Build economic models that include QC and QA cycle time as cost drivers. Recent automation analyses show cost and footprint reductions are feasible, but only when the full workflow—including analytics and quality review—is engineered as an integrated industrial system [267].

Chapter 11
Current Limitations and Unresolved Challenges

Abstract This chapter critically examines current limitations and unresolved challenges in MSC–EV research and translation. It highlights issues related to biological heterogeneity, standardization, dosing uncertainty, delivery barriers, and safety concerns. Quantitative evidence from recent studies is used to illustrate variability and reproducibility challenges. The chapter proposes research priorities and practical strategies to address these limitations and advance the field toward clinical maturity.

Keywords Limitations · Heterogeneity · Standardization · Dosing · Safety

Across the last decade, the EV therapeutics field has advanced fastest in proof-of-concept efficacy and engineering creativity, but translation continues to be constrained by a small set of recurring, tightly coupled limitations: (i) EV biological heterogeneity that is both real and method magnified; (ii) strong source dependence (donor, tissue, culture microenvironment, and media supplements) that can dominate variance in multi-omics and functional readouts; (iii) persistent standardization and metrology gaps in measurement, reporting, and reference materials; (iv) unresolved dose definition and scaling challenges; (v) systemic delivery barriers including rapid clearance and dominant uptake by the liver/spleen; (vi) practical instability including freeze–thaw losses and limited shelf-life understanding; and (vii) safety uncertainties—especially immunogenicity, thrombosis/procoagulant activity, and tumor interaction risks—amid still-heterogeneous clinical reporting [71] (Fig. 11.1).

Quantitatively, several "anchor" findings illustrate the scale of the problem space:

Reporting/Standardization EV-TRACK's EV-METRIC (a nine-component transparency score) was low overall in early literature and increased only modestly after guideline publication (mean ~19.8% pre-guidelines vs ~24.7% post-guidelines), with a larger improvement among studies that explicitly cited the guideline source (~35.8% vs ~22.4% non-citing) [268].

Inter-laboratory Reproducibility even when labs use aligned ultracentrifugation workflows, particle yield can vary substantially; one multicenter comparison reported yield variation with CV ~46.5% under the tested conditions, pointing to

N. M. Mussin, A. Tamadon, *Mesenchymal Stem Cells and Their Extracellular Vesicles*, https://doi.org/10.1007/978-3-032-36437-1_11

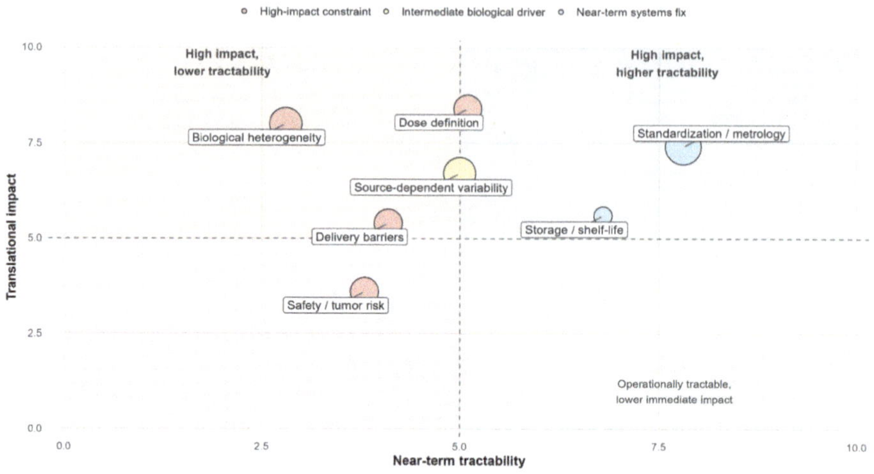

Fig. 11.1 Impact–tractability landscape of unresolved challenges in EV therapeutics. Major unresolved challenges in EV therapeutics positioned according to estimated translational impact and near-term tractability. Biological heterogeneity, dose definition, delivery barriers, and safety/tumor-related risks cluster in the high-impact but lower-tractability region, whereas standardization/metrology and storage control appear more tractable near-term priorities. Source-dependent variability occupies an intermediate position because it is biologically important but partially manageable through controlled sourcing and manufacturing strategies

equipment/operator effects and the need for calibration materials and tighter SOP control [269].

Dosing Dispersion meta-analysis of preclinical EV therapeutic studies identified up to ~3 orders of magnitude ("3 logs") variation in doses (whether expressed as protein mass or particle number), not resolved by stratifying by disease model, purification method, or source [270].

In Vivo Kinetics in animal studies summarized systematically, IV-administered small EVs showed rapid blood clearance—about 30% remaining at ~2 min, ~1.8% to 3.3% remaining by 5–30 min, and ~0% to 1.4% from 1 h onward; reported PK parameters spanned $T_{1/2}\alpha$ ~1.5 to 19.9 min and $T_{1/2}\beta$ ~34.6 to 184.5 min across a small number of comparable studies [271].

Storage Losses a 2024 systematic review found freeze–thaw cycles commonly reduce EV counts and RNA content; one cited study reported ~23% to 36% EV loss after a single freeze–thaw, ~37% to 43% loss after three cycles, and ~70% EV miRNA degradation after one freeze–thaw (study dependent) [153].

Clinical Safety Signal a meta-analysis of human EV therapy studies (335 patients across 21 reports) estimated serious adverse events (SAE) ~0.7% (95% CI 0.1–5.2%) and adverse events (AE) ~4.4% (95% CI 0.7–22.2%), but with notable heterogeneity in study design, manufacturing detail, and AE reporting; AE rates were higher in autologous vs allogeneic deployments and in engineered vs non-engineered products in subgroup analyses (interpretation limited by heterogeneity) [227].

At the clinical development landscape level, an analysis of the ClinicalTrials.gov registry reported 355 human-derived EV clinical trials initiated by the end of 2025, highlighting growth, but not yet resolution of the core CMC, dosing, and delivery constraints [228].

11.1 Evidence Synthesis Across Challenge Domains

11.1.1 Biological Heterogeneity

The dominant consensus view is that EVs represent molecularly and structurally heterogeneous populations; this heterogeneity is not a peripheral inconvenience but a core attribute that simultaneously enables broad function and complicates reproducible characterization and translation [71] (Fig. 11.2).

Heterogeneity is observable at multiple levels:

Physical scale: consensus documents describe EVs as nano- to micro-sized and emphasize that operational size categories lack strict cutoffs and can overlap substantially across isolation approaches. In MISEV2023, "small" is often used for EVs < ~200 nm, but the document explicitly warns there is no strict consensus on size cutoffs and many methods yield overlapping size profiles [71].

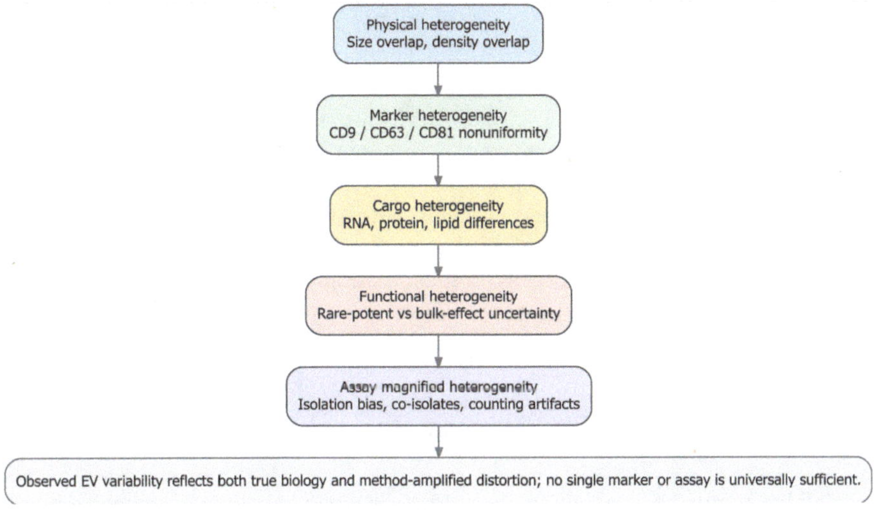

Fig. 11.2 Multilayer biological heterogeneity in therapeutic EV populations. Therapeutic EV heterogeneity is shown as a multilayer phenomenon spanning physical size, marker distribution, cargo composition, and biological activity. The figure emphasizes that heterogeneity is both intrinsic and method amplified, explaining why no single marker or assay can serve as a universal identity or quality surrogate

Marker expression and single-vesicle compositional variance: high-resolution imaging reviews summarizing single-EV measurements report variable copy numbers for canonical markers. One super-resolution/STORM-based example summarized in a 2023 review reported mean copy numbers on seminal EVs of CD63 ~ 12.6, CD81 ~ 1.6, and CD9 ~ 16.6, implying that "presence/absence" is not binary and that detectability can be marker and assay limited [272].

Subpopulation structure: single-EV profiling in plasma EVs has demonstrated that co-expression patterns of common tetraspanins are non-uniform; in one study, autoimmune disease plasma EVs showed significantly altered distributions of CD9-only, CD81-only, and CD9/CD81 double-positive EVs (with reported statistical differences such as $p \approx 0.04$ and $p \approx 0.002$ for specific comparisons) [273].

A key practical corollary is that no single marker (and no single technique) can be treated as "universal" for EV identity or for therapeutic quality assessment; even in highly cited quantification frameworks, authors explicitly note that none of the three canonical tetraspanins is present on every EV, motivating multi-marker approaches as imperfect but useful proxies [274].

11.1.2 Major Conflicting Findings and Likely Drivers

A central conflict in the literature is whether functional activity is dominated by a small number of potent subpopulations (rare but high-impact EVs) versus a broad bulk effect from major EV fractions. Many studies still rely on bulk assays that average over subtypes, which can make rare subpopulation dominance invisible or appear as "batch variability" [272]. A second conflict is definitional: some papers label their product "EVs" based on size and common markers, but consensus guidance discourages biogenesis-based naming without direct evidence, because mixed populations are typical and naming can overstate mechanism specificity [71]. Methodological drivers of apparent heterogeneity include co-isolation of non-vesicular particles (especially in biofluids), aggregation artifacts, and differences in particle counting/phenotyping technologies, several of which cannot discriminate protein/lipoprotein particles from EVs [274].

11.1.3 Key Unresolved Questions

1. Which EV attributes (size fraction, surface phenotype, cargo signatures) are causal versus merely correlated with therapeutic activity across indications [71]?

2. What fraction of "therapeutic effect" in published studies is attributable to EVs versus co-isolated non-EV material (protein aggregates, lipoproteins, soluble mediators) [274]?
3. Can single-EV measurements be standardized enough to serve as a release test backbone, rather than an exploratory characterization layer [272]?

11.1.4 Research Priorities and Practical Recommendations

Research priorities should shift from "more markers" to mechanistically anchored heterogeneity mapping: link single-EV phenotypes and cargo to functional assays, with explicit negative controls for non-EV components as emphasized in community guidelines [71]. Practically, for a therapeutic developer operating without a single target indication, the least-regret approach is to define a multidimensional identity profile: (i) operational size/density window, (ii) multi-marker protein profile (positive and negative), (iii) purity indicators, and (iv) a potency assay that is stable to modest heterogeneity but sensitive to key functional cargo [71].

11.2 Source-Dependent Variability

11.2.1 Current Evidence and Quantitative Ranges

Evidence across 2016–2026 consistently indicates that EV composition and function depend strongly on parent cell/tissue source, donor, and culture microenvironment. A clear demonstration of donor-linked variance comes from quantitative proteomics performed on EVs from multiple adipose MSC donors: principal component analysis separated samples primarily by donor origin, with the first principal component explaining ~57% of variance and the second ~31%, while replicates within the same batch clustered tightly—i.e., high intra-batch consistency but clear inter-donor structure [275].

Notably, functional outcomes may display less donor divergence than omics. In the same donor comparison study, multiple batches from three MSC donors showed comparable anti-inflammatory effects in an in vitro inflammatory model (TNF-α and IL-6 readouts reported across concentrations). This pattern—omics variability coexisting with partially conserved function—appears repeatedly and creates practical uncertainty about how much donor selection matters for therapeutic efficacy versus for safety or off-target risk [275].

Source variability is also evident across tissue-resident MSC types. A comparative analysis of EVs isolated from endometrium-, placenta-, and dental pulp–derived MSCs reported substantial differences in physical heterogeneity, with examples of PDI ~0.51 ± 0.04 and 0.61 ± 0.18 for two EV sets, interpreted by the authors as suggesting mixed EV types/apoptotic bodies based on size heterogeneity, and broader differences in surface marker profiles and physicochemical signatures [126].

Culture configuration can measurably reshape EV cargo resemblance to in vivo EVs. In one study comparing 2D vs 3D culture in a cancer model, the 3D culture EV small RNA profile showed ~96% similarity to in vivo circulating EVs from patient plasma, suggesting 3D contexts may produce more "in vivo-like" EV RNA patterns than standard monolayers (at least in that model) [276].

Media supplementation is another major source-dependent driver. A 2025 ACS nano-study (as summarized in its indexed abstract) reported that human platelet lysate (HPL) supplementation increased MSC–EV production without compromising "critical quality attributes" and reported "stability" of HPL-conditioned EVs across UC-MSC donors via multi-omics, while also indicating global functional alterations under different culture conditions [277]. The field simultaneously recognizes that supplements can introduce exogenous particle contamination, motivating methods to generate EV-free supplements; one study described a scalable TFF approach to produce EV-free HPL (and EV-free FCS) and reported that the EV-free additive supported MSC culture while enabling higher EV production of constant quality over time [278].

11.2.2 Major Conflicting Findings and Likely Drivers

A persistent conflict is whether donor variability is a "feature" (allowing personalization) or a critical risk (undermining comparability and regulatory reproducibility) (Fig. 11.3). Proteomics often shows strong donor clustering, but functional assays can show partial convergence; which level dominates likely depends on indication and on whether the intended mechanism is broad immunomodulation versus a narrow, pathway-specific effect [275].

Another conflict is how to interpret source-driven tropism. Some reviews note intrinsic liver tropism for some MSC-EVs, framing it as an advantage for hepatic indications and a barrier elsewhere, but evidence varies with labeling methods, administration route, and EV engineering, making "tropism" itself a measurement-sensitive conclusion [271].

11.2.3 Key Unresolved Questions

1. What is the optimal choice between single-donor, multi-donor pooling, and cell line/immortalized producer systems, when balancing potency, safety, and consistency [275]?

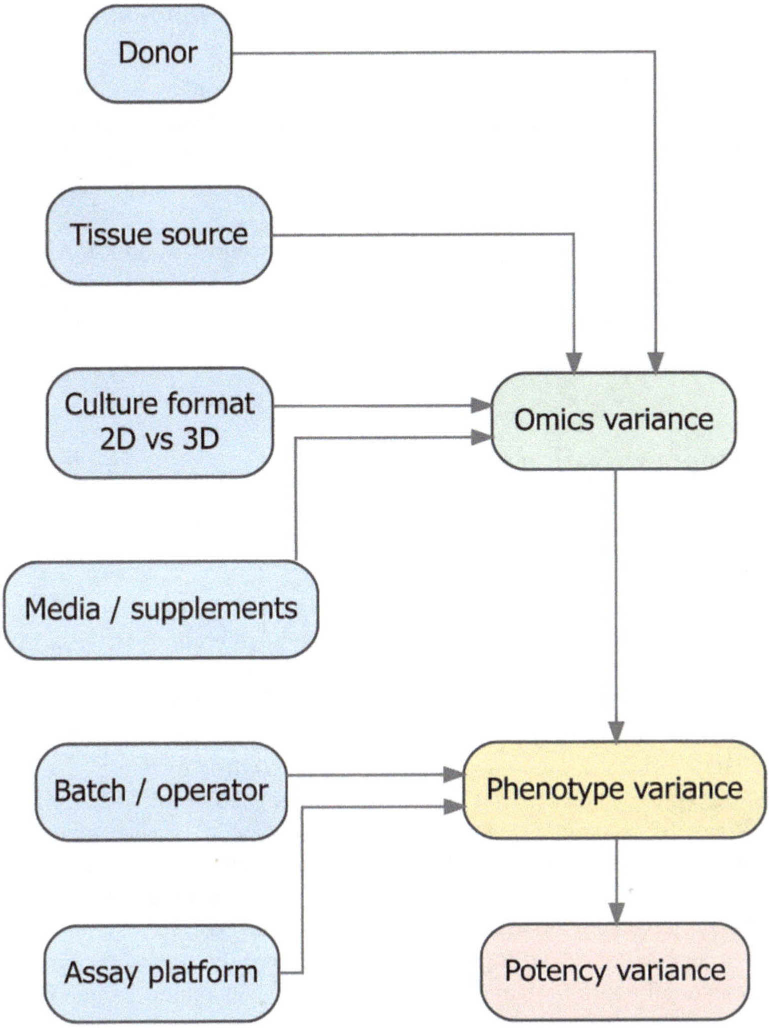

Fig. 11.3 Source-dependent variance architecture in EV production and function. Donor origin, tissue source, culture configuration, media supplementation, batch effects, and assay platform each contribute to observed EV variance, but not always at the same biological level. Omics profiles may diverge strongly, while some functional outputs remain partially conserved, creating uncertainty about which source-driven differences are therapeutically meaningful

2. Which culture parameters (2D/3D, oxygen tension, media supplements) drive functionally meaningful changes versus "omics-detectable but potency-neutral" changes [276]?
3. How should "in vivo-like" cargo resemblance be operationalized for therapy [276]?

11.2.4 Research Priorities and Practical Recommendations

A practical research priority is to design donor/culture studies with variance partitioning: quantify how much variance is driven by donor, batch, operator, and assay (rather than reporting only mean differences). The donor proteomic PCA example (57% of variance linked to donor) illustrates that such decomposition is achievable and informative [275].

For practical development, recommend adopting "source-aware CQAs": define a core identity signature expected to be conserved across donors (or a controlled donor pool), alongside a set of variable attributes that are monitored as risk indicators [23].

11.3 Standardization Challenges

11.3.1 Current Evidence and Quantitative Ranges

The strongest evidence of standardization gaps appears in three domains: reporting transparency, measurement comparability, and reference materials/controls (Table 11.1).

Community consensus and uptake: MISEV2023 was built through large-scale community feedback (survey circulated to ~5700 EV researchers with 1025 responses) and includes formal consensus metrics for sections; for example, Section 3 of MISEV2023 (pre-analytical variables through storage) had ~70.4% "completely" and ~28.5% "mostly" agreement, while Section 6 of MISEV2023 (technique-specific reporting considerations) had ~70.6% "completely" and ~27.5% "mostly" agreement. This quantifies broad agreement on what should be reported and controlled [71].

However, transparent reporting remains inconsistent. EV-TRACK's EV-METRIC analysis (nine essential reporting components) found only modest overall improvement after guideline publication: mean EV-METRIC increased from ~19.8% pre-guidelines to ~24.7% post-guidelines, with a stronger increase for studies citing the guideline (~35.8% vs ~22.4% non-citing) [268].

Measurement comparability gaps are magnified by inter-laboratory variability. In an inter-lab ultracentrifugation transferability study, particle yield variation in one round showed CV ~46.51%, attributed to operator/equipment effects under the tested conditions and motivating reference material use to calibrate instruments and benchmark methods [269].

Method-specific limitations: several widely used physical counting methods do not distinguish EVs from common confounders. A rapid comparison framework using ultrasensitive immunoassays emphasizes that NTA/DLS/TRPS-style approaches provide size and concentration but may be unable to discriminate EVs from similarly sized lipoproteins and protein aggregates, motivating multi-marker immuno-quantification and paired contaminant measurement for purity/yield comparisons [274].

Table 11.1 A translation-oriented comparison of commonly used separation/purification approaches, emphasizing standardization risks

Workflow element	Typical methods	Strengths for translation	Standardization risks and common confounders	Practical "minimum controls" to report/ implement
Primary concentration	Differential UC; ultrafiltration; TFF	TFF supports scale-up; UC is widespread and accessible	UC is operator/ equipment sensitive; yield can vary strongly across labs (CV ~46.5% reported in multicenter UC comparison conditions) [269]	Record rotor type/k-factor equivalents; include reference particles/ spike-ins where feasible; track yield/purity across steps [269]
Size-based separation	SEC	Can improve purity vs crude UC in biofluids and conditioned media; scalable with automation	Fraction selection and column variability; still cannot on its own prove EV identity	Report fraction ranges; quantify multiple EV markers and a contaminant marker to compute relative purity/yield [274]
Precipitation	PEG/ commercial kits	Convenience and throughput	High co-precipitation risk (proteins/ lipoproteins); cross-study purity not comparable	Require orthogonal validation and contaminant quantification; avoid treating particle counts as EV counts without marker evidence [274]
Affinity capture	Antibody/ ligand capture	Potentially enriches specific subpopulations	Captures only marker-positive subfractions (bias); reagent specificity issues	Establish reagent specificity; report epitope/marker rationale; include non-EV controls [71]
Integrated workflows	TFF → SEC; multi-step manufacturing	Aligns with industrial and GMP needs; enables batch processing	More steps can introduce loss and variability if not controlled	In-process controls at each stage; define CQAs and comparability strategy across batches [280]

Finally, systematic reviews of EV drug delivery studies emphasize a "gold standard" gap in purity determination: impurities such as protein aggregates or LDL/VLDL-like particles can be difficult to discriminate, and the literature has not converged on a single purity metric; reviews recommend orthogonal combinations [279].

11.3.2 Major Conflicting Findings and Likely Drivers

A major conflict is whether discrepancies across studies reflect "real biology" or "methodology." For example, EV tropism conclusions are robust at a high level, yet relative organ signals can vary with labeling modality, imaging depth limitations, and whether the signal tracks the EV membrane, its cargo, or a dissociated label. MISEV2023 explicitly cautions that labeling can affect biodistribution and function, and systematic reviews identify differences between in situ and ex vivo biodistribution estimates [71].

A second conflict concerns what constitutes "sufficient characterization." MISEV2023 quantifies strong community agreement on technique-specific reporting, yet many studies remain underreported by EV-METRIC criteria; this likely reflects friction costs (time, expertise, and instrument access) and the lack of universally adopted reference materials or shared assay panels [71].

11.3.3 Key Unresolved Questions

1. What is the minimal, regulator-credible panel of CQAs (identity, purity, potency, safety) that can be applied across EV sources and indications [280]?
2. Which reference materials (biological or synthetic) best enable cross-platform calibration for heterogeneous EV samples (beyond monodisperse bead standards) [269]?
3. How can reporting standards be made "default" rather than optional [268]?

11.3.4 Research Priorities and Practical Recommendations

A near-term research priority is inter-laboratory round-robin studies spanning multiple isolation methods and readouts (yield, purity, potency), explicitly partitioning variance into intra-operator, intra-lab, and inter-lab components—an expansion that multicenter UC studies themselves recommend to interpret CV values meaningfully [269] (Fig. 11.4). Practically, developers should adopt MISEV2023 as a reporting and experimental design checklist and embed an EV-TRACK-style scorecard mentality into internal SOPs, even if external reporting remains variable; the numerical EV-METRIC deltas show that such checklists measurably improve transparency [268].

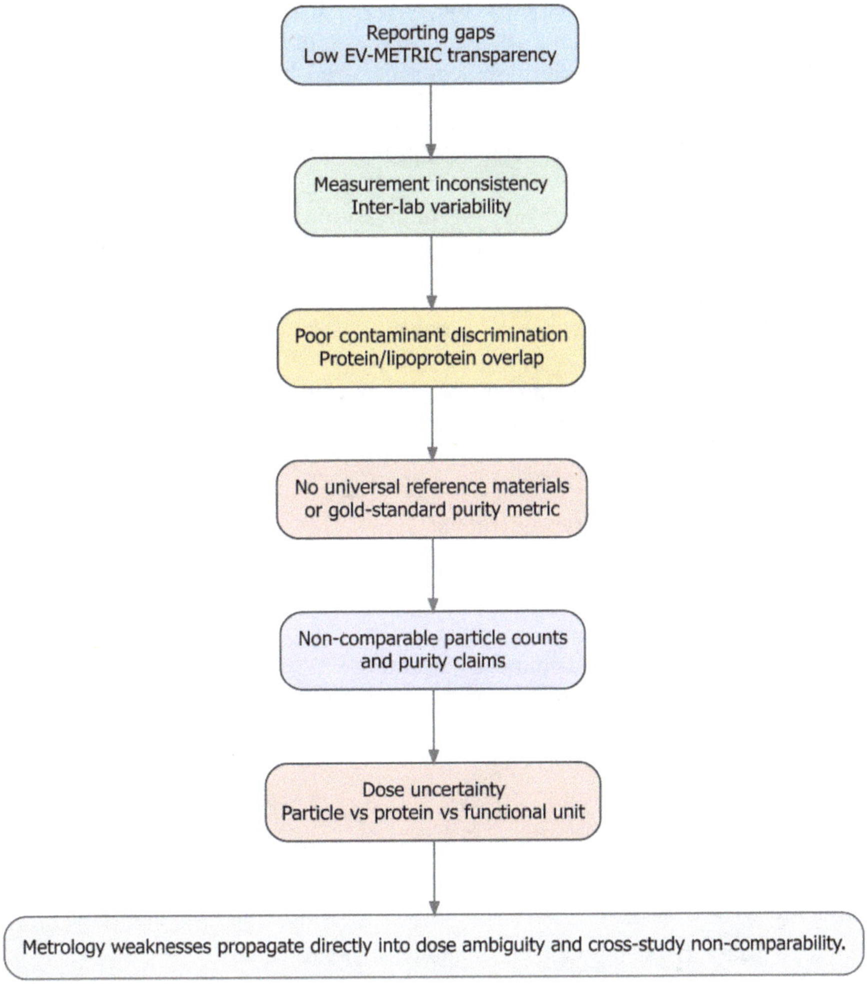

Fig. 11.4 From metrology gaps to dose uncertainty: the EV measurement problem. Reporting gaps, inter-laboratory variability, incomplete contaminant discrimination, and lack of universal reference materials are linked here as a single causal chain that destabilizes dose interpretation. The figure shows how metrology weaknesses propagate directly into cross-study noncomparability and persistent uncertainty about what an EV dose actually represents

11.3.5 Dosing Uncertainty

11.3.5.1 Current Evidence and Quantitative Ranges

Dosing is constrained by three interlocking uncertainties: (i) what the "active ingredient" is (particle, cargo, protein subset, functional unit), (ii) which quantification method is trustworthy for the matrix, and (iii) how to scale dose across species, routes, and indications.

Preclinical Heterogeneity Is Extreme A 2023 review reflecting on a meta-analysis of 64 preclinical therapeutic EV studies reported dose variation up to ~3 logs (protein based or particle based), not resolved by disease type, purification method, or cell source [270].

In humans, a 2024 scoping review compiling 40 published human studies illustrates that dosing units vary widely even within similar product classes: particles/kg, particles per dose, particles per mL, mL of an EV product, and topical mass-based dosing all appear in the literature. Representative examples drawn from the scoping review include:

- IV "three doses on three consecutive days" at ~0.8 to 1.2×10^9 particles/kg (case report context) [281].
- A dose comparison trial using ~1.2×10^{12} vs ~0.9×10^{12} EV particles per dose (high vs low dose) via IV infusion [281].
- IV administration of 1 mL of a nanoparticle/EV biologic with reported concentrations around ~3.26×10^{11} particles/mL in case reports and ~1 to 5×10^{11} particles/mL in a proof-of-concept clinical context [281].
- Nebulized/inhaled dose escalation ranges reported as ~2×10^8 to 16×10^8 particles and an inhalation regimen of ~2×10^8 EVs daily for 5 days (study dependent) [281].

MISEV2023 explicitly recommends reporting EV/recipient ratios and the physiological relevance of delivered dose, underscoring that dose reporting is not merely "documentation hygiene" but a necessary condition for mechanistic interpretation and inter-study comparability [71].

11.3.6 Major Conflicting Findings and Likely Drivers

Conflict appears in whether dosing should be standardized to:

- Particle number (intuitive for "nanoparticle-like" vectors but sensitive to counting method and contaminants).
- Total protein mass (easy but nonspecific).
- "Cell equivalents" (ties to manufacturing but not necessarily to functional units)
- Potency-normalized functional units (conceptually strongest but currently weakly standardized) [270].

Another conflict is the degree to which dosing choices account for EV pharmacokinetics and biodistribution. Reviews argue that doses are sometimes selected without incorporating published PK/BD patterns, which is especially problematic given rapid clearance and liver/spleen sequestration after IV dosing [270].

11.3.6.1 Key Unresolved Questions

1. What dosing metric best predicts efficacy and safety across indications: particle number, cargo amount, or potency-defined functional units [270]?
2. How should dose be scaled across routes (IV vs local vs inhaled vs intranasal) when biodistribution and clearance differ substantially by route [271]?
3. How should repeated dosing be designed given potential immunogenicity/antibody development and the difficulty of measuring true in vivo exposure ("toxicokinetics" for heterogeneous EV mixtures) [23]?

11.3.6.2 Research Priorities and Practical Recommendations

A high-priority research target is potency assay development that can (i) be run as a release test, (ii) correlate with in vivo efficacy, and (iii) remain robust to reasonable biological heterogeneity. Community guidance supports shifting emphasis toward functional readouts and carefully defined negative controls rather than relying solely on particle counts [270]. Practically, developers should report dose in at least two orthogonal units and explicitly document the particle counting method and its contamination limitations. This dual reporting supports later harmonization even if the field converges on a single preferred unit [274].

11.3.7 Delivery Barriers

11.3.7.1 Current Evidence and Quantitative Ranges

Systemic delivery of EVs faces the classic barriers of nanoparticle therapeutics—rapid clearance, reticuloendothelial uptake, and limited accumulation at many target tissues—but with additional complexity: EVs are biological particles whose uptake can be active, source dependent, and labeling sensitive [271].

A systematic review of animal biodistribution studies provides one of the clearest quantitative summaries available: after IV administration of small EVs, ~30% of injected dose remained in circulation at ~2 min, dropping to ~1.8% to 3.3% by 5–30 min and ~0% to 1.4% from 1 h onward; in three cited studies, ~95% was removed within 5 min. Reported PK ranges across a limited number of comparable studies included $T_{1/2}\alpha$ ~1.5 to 19.9 min and $T_{1/2}\beta$ ~34.6 to 184.5 min, with AUC and MRT ranges also reported (AUC ~0.7% to 3.2% dose·h/mL, MRT ~0.5 to 0.847 h) [271].

The same systematic review indicates that, across time windows, apparent distribution differs by EV size class and organ: it reports low to no localization of large EVs in the brain and heart across assessed periods and dynamic shifts between lung and liver signals, underscoring how size fractions intersect with first-pass pulmonary capillary beds and subsequent clearance pathways [271].

A 2024 radiolabel/SPECT study of human UCMSC-EVs further quantifies rapid kinetics: it reported a distribution phase of ~0.85 ± 0.28 min and an elimination phase of ~25.22 ± 20.76 min in mice, with major uptake in the liver and spleen and renal excretion suggested by bladder signal [282].

Importantly, many preclinical EV delivery studies still do not report biodistribution. A systematic review of EVs as drug delivery systems found that only ~49.7% of included preclinical studies performed biodistribution experiments, despite biodistribution being essential for interpreting targeting and toxicity, and IV administration dominated routes (~63.1% overall, rising to ~88.2% in 2020) [279].

11.3.7.2 Major Conflicting Findings and Likely Drivers

A recurring conflict is the magnitude of lung vs liver/spleen localization, which systematic reviewers suggest can differ between in situ optical imaging and ex vivo organ analysis [271]. Another conflict is label fidelity. Consensus guidance cautions that labeling may itself alter biodistribution, pharmacokinetics, and function and recommends controlling for these effects; without such controls, studies may measure the fate of labels rather than EVs (or may measure both but without disambiguation) [71].

11.3.7.3 Key Unresolved Questions

1. What proportion of systemic dose can realistically reach specific organs without engineering, and what is the ceiling imposed by mononuclear phagocyte clearance [271]?
2. Which delivery strategies (local delivery, intranasal/inhaled, biomaterial depots, surface engineering) best improve target exposure without introducing new safety liabilities [279]?
3. How should "targeting success" be defined—by EV membrane accumulation, cargo delivery, functional response, or all three [71]?

11.3.7.4 Research Priorities and Practical Recommendations

A priority is to standardize PK/BD study design for EVs: define what is measured (whole EV vs cargo), use at least two tracking modalities when feasible, and incorporate controls for label dissociation, as recommended in consensus guidance and drug delivery systematic reviews [71] (Fig. 11.5). Practically, absent a specific indication, a conservative strategy is to treat IV EV delivery as primarily suited for organs with high first-pass or clearance exposure (liver/spleen/lung) unless robust targeting evidence exists and to prioritize local or route-matched delivery when feasible. Human dosing examples in the scoping review indicate that inhaled/

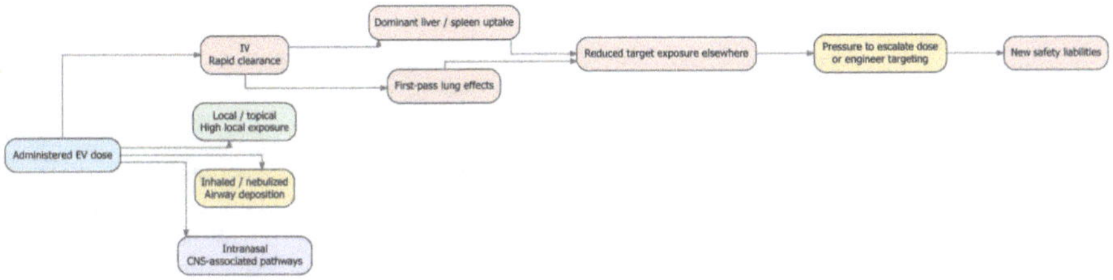

Fig. 11.5 Exposure bottlenecks after EV administration: clearance, sequestration, and target shortfall. Intravenous EV delivery is dominated by rapid clearance and reticuloendothelial uptake, especially in the liver and spleen, resulting in reduced target tissue exposure elsewhere. In contrast, local, inhaled, and intranasal routes may improve route-matched exposure but introduce distinct design constraints; poor target access then drives dose escalation and targeting–engineering pressure with potential safety consequences

nebulized regimens often operate in ~10^8 particle ranges, distinct from IV particle/kg regimens, reinforcing the route dependence of dose design [271].

11.3.8 *Storage and Shelf-Life Issues*

11.3.8.1 Current Evidence and Quantitative Ranges

EV stability is now recognized as a translation-critical variable, not a logistical afterthought; EVs' nanoscale structure makes them sensitive to temperature shifts, ice crystal formation, aggregation, and cargo degradation [153] (Table 11.2). A 2024 systematic review of storage/freezing protocols (50 included studies) concluded that rapid freezing and maintaining constant subzero temperatures (optimally −80 °C) best preserve EV quantity and cargo, while repeated freeze–thaw cycles reduce particle concentration and RNA content and impair bioactivity, often with increased size/aggregation and membrane deformation seen on microscopy [153].

Critically, this review provides quantitative examples (study dependent):

- ~23% to 36% EV loss after a single freeze–thaw cycle (blood cell–derived EVs in one cited study) [153]
- ~37% to 43% reduction in EV quantity after three freeze–thaw cycles (in one cited report) [153]
- ~70% EV miRNA degradation after a single freeze–thaw cycle in one report [153]

The same review highlights stabilizers: trehalose and related cryoprotectants can help preserve integrity; it also notes that storage in native biofluids can improve stability relative to purified EVs in buffer (context dependent). Long-duration data exist but remain sparse; one cited study reported that 2 years at −80 °C in PBS supplemented with human serum albumin and trehalose resulted in minimal RNA/protein degradation compared with PBS alone [153].

Table 11.2 Comparing storage conditions [153]

Condition	What is observed in recent evidence	Quantitative examples (study dependent)	Practical implication
Constant −80 °C (rapid freezing preferred)	Better particle/cargo preservation than higher temperatures; reduces aggregation risk vs fluctuating conditions	Recommended as "optimal" subzero in a 50-study systematic review	Treat as baseline for research comparability; validate for clinical feasibility
Freeze–thaw cycling	Decreases particle concentration and RNA; increases size/aggregation; can impair bioactivity	~23–36% EV loss after 1 cycle; ~37% to 43% loss after 3 cycles; ~70% miRNA degradation after 1 cycle (specific reports)	Aliquot; record freeze–thaw count; avoid comparing samples with different histories
Room temperature/4 °C for extended periods	Can lead to degraded/empty EVs and marker loss in some studies	Example: Marker detectability declines rapidly at RT in some reports summarized by the systematic review	Avoid for anything beyond short handling windows unless validated
Stabilizers	Help maintain integrity; can reduce aggregation; may support longer storage	Two-year −80 °C storage with albumin + trehalose minimal degradation vs PBS alone (one study cited)	Formulation is part of the "product"; must be locked early and carried through comparability

MISEV2023 explicitly recommends minimizing freeze–thaw cycles and cautions that samples with different numbers of freeze cycles may not be directly comparable—aligning consensus guidance with observed losses [71].

11.3.8.2 Major Conflicting Findings and Likely Drivers

Conflicting findings arise around whether one freeze–thaw cycle is "acceptable." The systematic review notes some studies report minimal effects after one cycle, while others report measurable losses after one; many studies show clearer declines after two or more cycles. This conflict likely reflects differences in EV source, formulation buffer, cryoprotectant use, freezing/thawing rates, measurement methods, and whether stability is judged by particle count versus potency [153].

A second conflict is the true feasibility of pharmaceutical shelf-life. Some reviews state no robust shelf-life data exist for exosome formulations, and clinical-grade storage at −80 °C is operationally impractical for many supply chains, motivating lyophilization and stabilizer development but with limited standardized comparability [283].

11.3.8.3 Key Unresolved Questions

1. What are the most predictive stability-indicating assays for EV products: particle count, marker panel, cargo integrity, or potency [153]?
2. Can lyophilization (with stabilizers) yield room-temperature shelf-life without compromising targeting/uptake or introducing aggregates that mimic "dose increases" [153]?
3. What are acceptable shipping and handling thresholds (temperature excursions, vibration, container adsorption) for clinical distribution [153]?

11.3.8.4 Research Priorities and Practical Recommendations

A near-term research priority is to design ICH-style stability protocols for EVs (even before full regulatory harmonization): defined storage temperatures, excursion studies, agitation/shipping simulation, container closure effects, and potency correlation. Practically, minimum recommendations supported by current evidence are as follows: (i) store at −80 °C when feasible, (ii) aliquot to avoid repeat freeze–thaw, (iii) evaluate cryoprotectants such as trehalose where compatible with intended route, and (iv) treat "particle loss" and "functional loss" as distinct endpoints requiring different assays [153].

11.3.9 Safety Concerns and Tumor-Related Questions

11.3.9.1 Current Evidence and Quantitative Ranges

Human safety evidence is growing but remains heterogeneous in quality and reporting. A 2024 systematic review/meta-analysis of clinical trials assessing EV therapy included 21 reports (335 patients) and estimated overall SAE frequency ~0.7% and AE frequency ~4.4%, with low heterogeneity for SAE (I^2 ~ 0%) but substantial heterogeneity for AE (I^2 ~ 60%). The most frequent SAE was liver dysfunction ($n = 2$ in the dataset), with other single occurrences reported. Subgroup analyses found no significant difference in SAE between autologous and allogeneic administration or engineered vs non-engineered products, but higher AE in autologous vs allogeneic and in engineered vs non-engineered subgroups; these subgroup findings require caution because AE reporting and study designs varied widely [227].

A 2024 quality/safety perspective on therapeutic EV products emphasizes a case-by-case nonclinical safety strategy: evaluate on-target vs off-target toxicity, consider immunogenicity (especially in allogeneic settings where MHC molecules may be present), and acknowledge that toxicokinetics is difficult because EV preparations are heterogeneous and not easily quantifiable as discrete species. The same paper flags carcinogenicity concerns in principle—particularly for EVs derived from immortalized cells and the possibility of retrotransposon sequences—and it argues that FBS should be avoided when possible as it can introduce confounding/contaminating biological materials [23].

Thrombosis/procoagulant risk is a specific safety concern because EVs can carry phosphatidylserine and tissue factor, and reviews on tissue factor-bearing EVs and EVs in thrombosis frame EVs as potential mediators of coagulation activation depending on source and context; this becomes translation relevant when EV products are administered systemically or repeatedly [269].

11.3.9.2 Tumor-Related Questions: Current Evidence and Conflicts

Tumor risk is not limited to "tumor-derived EVs"; rather, the concern is that EVs from multiple sources may interact with tumor biology (microenvironmental modulation, immune regulation, angiogenesis). Reviews describe MSCs and their secretome as potentially tumor promoting in some contexts, emphasizing that heterogeneity of MSC sources, isolation methods, and tumor subtype differences complicate interpretation [167]. Simultaneously, other reviews highlight anti-tumor mechanisms of MSCs/MSC-derived components in specific models, and literature on MSC-related EVs in cancers such as osteosarcoma explicitly frames the evidence as mixed and context dependent, with both tumor-suppressive and tumor-supportive pathways discussed across studies [284]. The practical translation problem is that "tumor interaction" is often not a primary endpoint in non-oncology EV trials, so low SAE rates do not fully resolve long-term carcinogenicity or tumor promotion risks, especially for chronic indications requiring repeated dosing [227].

A recent systematic review and meta-analysis of in vitro breast cancer studies by Mussin et al. [285] further illustrates the context-dependent relationship between MSC–EVs and malignant cells. This evidence strengthens the conclusion that regenerative EV products cannot automatically be considered oncologically neutral. Producer cell source, conditioning state, EV cargo, dose, exposure duration, cancer subtype, and recipient cell phenotype should therefore be incorporated into product-specific tumor interaction testing.

11.3.9.3 Key Unresolved Questions

1. What is the long-term risk profile of repeated EV dosing (immunogenicity, antibody development, attenuation of effect, hypersensitivity), especially for engineered EVs with modified surfaces/cargo [23]?
2. Which safety assays should be mandatory release tests (sterility/endotoxin, procoagulant activity, residual DNA/RNA risk markers, oncogenic cargo screening), and which are indication specific [23]?
3. How should tumor promotion risk be assessed for EVs intended for immunosuppression or tissue repair, particularly in populations with elevated baseline cancer risk [167]?

11.3.9.4 Research Priorities and Practical Recommendations

A field-level research priority is to define safety-relevant CQAs linked to plausible hazards: (i) immunogenic surface molecules, (ii) procoagulant activity (tissue factor/PS exposure), (iii) residual nucleic acid risks (including retrotransposon activity concerns in immortalized/unknown cell sources), and (iv) off-target immunosuppression or immune activation, depending on intended mechanism [23]. Practically, a conservative "translation-ready" safety package should include standardized AE/SAE reporting aligned to trial norms, explicit tumor surveillance plans for chronic dosing where plausible, and route-specific safety monitoring. The low pooled SAE rate in current meta-analyses is encouraging, but it is not a substitute for harmonized reporting and longer follow-up in controlled trials [227].

11.4 Cross-Cutting Research Priorities and Practical Recommendations

A coherent research agenda for EV therapeutics in an unspecified therapeutic context can be framed as a "translation stack," where progress at higher levels depends on solving lower-layer constraints (Fig. 11.6):

1. Metrology and reference materials first: without broadly accepted reference materials and cross-platform calibration approaches, inter-laboratory comparability will remain limited; multicenter UC studies explicitly anticipate reference materials to calibrate and benchmark methods, and MISEV2023 consensus highlights technique-specific reporting as a high-agreement domain [269].
2. Purity as a functional variable: systematic reviews note the lack of a gold standard for purity and the inability of some particle counting methods to discriminate contaminants; translation efforts should treat purity measurement as mechanistic (what is delivering the effect?) rather than cosmetic [274].
3. Potency-anchored dosing: the persistent 3-log dispersion in preclinical dosing argues that particle/protein units alone are insufficient; dose definition should be anchored to reproducible potency assays—ideally ones that can be linked to exposure and outcome [270].
4. Route-matched delivery development: systemic IV delivery faces rapid clearance and dominant liver/spleen uptake; route selection and delivery engineering should be treated as first-order design variables, not secondary decisions [271].
5. Stability-indicating shelf-life programs: freeze–thaw losses and cargo degradation are large enough to plausibly confound dose–response and safety; stability-indicating assays and handling constraints must be built into CMC early [153].
6. Risk-based safety and tumor interaction assessment: pooled SAE rates are low, but mechanistic risks (immunogenicity, coagulation activation, oncogenic cargo transfer in principle) warrant structured, product-specific safety packages and harmonized clinical reporting [227].

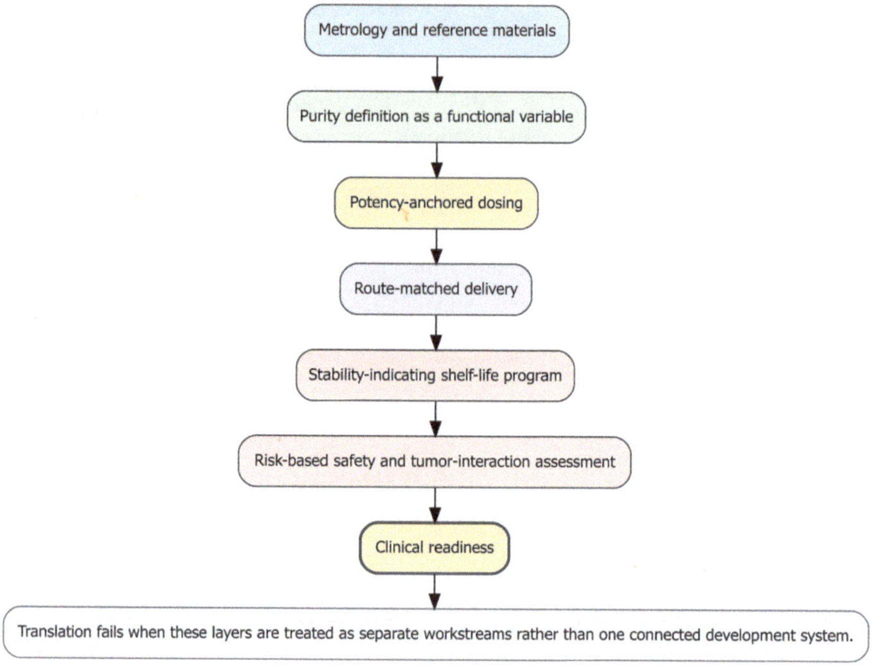

Fig. 11.6 Translation stack for EV therapeutics: stability, safety, and clinical readiness. This figure synthesizes Chap. 11 into a translation stack in which higher-level clinical readiness depends on solving lower-layer technical and biological constraints. Metrology and purity form the foundation, potency-anchored dosing and route-matched delivery define usable exposure, stability preserves the intended product over time, and risk-based safety assessment determines whether repeated clinical use is defensible

Chapter 12
Future Perspectives in MSC–EV-Based Cell-Free Therapy

Abstract This chapter explores future perspectives in MSC–EV-based cell-free therapy, focusing on precision medicine, bioengineering, and integration with artificial intelligence and omics technologies. It discusses emerging strategies for personalized therapy, advanced manufacturing, and clinical adoption. The chapter outlines a roadmap for transitioning MSC–EVs into programmable biologics, emphasizing the importance of standardization, potency definition, and regulatory alignment for future clinical success.

Keywords Future perspectives · Precision medicine · EV engineering · AI · Regenerative therapy

MSC–EVs are moving from "promising paracrine effectors" to "engineerable biologic-like therapeutics" as the field converges on three enabling capabilities: (i) rigorous product definition and quality frameworks (updated minimal information standards, trial landscape analyses, and medicinal product development guidance), (ii) scalable biomanufacturing and analytics (single-vesicle and calibrated counting methods, potency assay frameworks, and route-aware nonclinical packages), and (iii) programmable engineering (cell/EV surface re-targeting, nucleic acid and protein payloading, and controlled-release formats) [71].

Clinical evidence is no longer limited to anecdotal reports: multiple human studies have tested MSC–EV products across respiratory, inflammatory, and regenerative indications, including (i) nebulized clinical-grade MSC–EVs with healthy-volunteer safety data and preclinical efficacy (Shi et al., 2021; trial registration NCT04313647), (ii) randomized placebo-controlled EV infusion in COVID-19 respiratory failure (NCT04493242) [161], and (iii) first-in-human engineered EVs delivering KRAS-targeting siRNA (NCT03608631) [163, 286].

A persistent bottleneck is that EV "identity" does not guarantee "activity": donor-to-donor and lot-to-lot variability can occur despite standardized manufacturing, and surrogate markers may not correlate with immunomodulatory potency. Future adoption therefore hinges on mechanism-aligned, fit-for-purpose potency assays, validated dose metrics, and comparability strategies that keep pace with engineering complexity [287] (Fig. 12.1).

N. M. Mussin, A. Tamadon, *Mesenchymal Stem Cells and Their Extracellular Vesicles*, https://doi.org/10.1007/978-3-032-36437-1_12

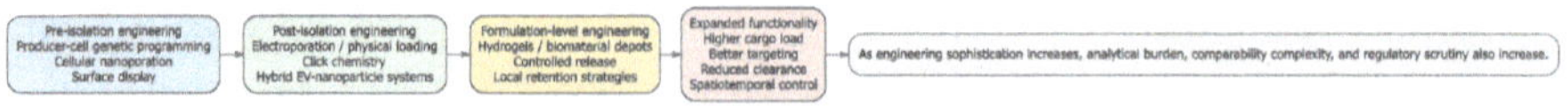

Fig. 12.1 Engineering design space for MSC–EV therapeutics: from producer cell programming to controlled-release formulations. This figure maps the engineering design space of MSC–EV therapeutics across pre-isolation, post-isolation, and formulation-level interventions. Pre-isolation strategies include donor cell genetic programming, cellular nanoporation, and surface display engineering; post-isolation approaches include electroporation, Click Chemistry functionalization, and hybrid EV–nanoparticle systems; formulation-level strategies include hydrogels and biomaterial depots for spatiotemporal control. The figure highlights how engineering expands functional capability but simultaneously increases analytical, comparability, and regulatory complexity

12.1 Precision EV Therapeutics

State of the art. "Precision" in MSC–EV therapeutics now increasingly means precision in product definition, measurement, and mechanism alignment, not only disease targeting. Updated minimal information standards (MISEV2018 → MISEV2023) clarify nomenclature, emphasize separation/characterization transparency, and caution against using biogenesis-based terms unless specifically demonstrated; they also foreground advanced methods that are increasingly relevant for therapeutics (single-vesicle analytics, reference materials, and functional evidence) [5] (Fig. 12.2). A parallel "precision" layer has emerged for MSC–sEV identity metrics, defining quantifiable attributes for cellular origin, lipid membrane vesicles, and integrity as a basis for data comparability and eventual release criteria [222].

Key recent advances. A central advance for precision is the shift from bulk averages (size/NTA and total protein) toward heterogeneity-resolving analytics (single EV profiling and calibrated counting). Reviews and method studies emphasize that single-vesicle approaches can serve both discovery and quality control by quantifying EV subpopulations and reducing ambiguity about "what is in the vial" [288].

Another precision enabler is the move toward route-aware pharmacology and modeling. EV pharmacokinetics/biodistribution reviews and PBPK modeling studies highlight that route (IV vs inhaled vs local) and labeling/measurement limitations strongly shape exposure and interpretation, making route-specific development essential for dose selection and risk mitigation [270].

Technical challenges. The first technical challenge is measurement truthfulness: EV preparations are heterogeneous and can be confounded by co-isolated non-EV particles, while size and concentration outputs depend on the platform and sample preparation. MISEV2023 explicitly notes that "size- or density-defined" labels are method dependent and that rigorous separation/characterization choices must be reported and justified [71].

A second challenge is functional delivery efficiency. Quantitative cellular studies show that uptake and cytosolic content release can be low yield and dependent on endosomal processes; these constraints directly limit precision claims about "payload delivered per dose" unless measured with validated assays [289].

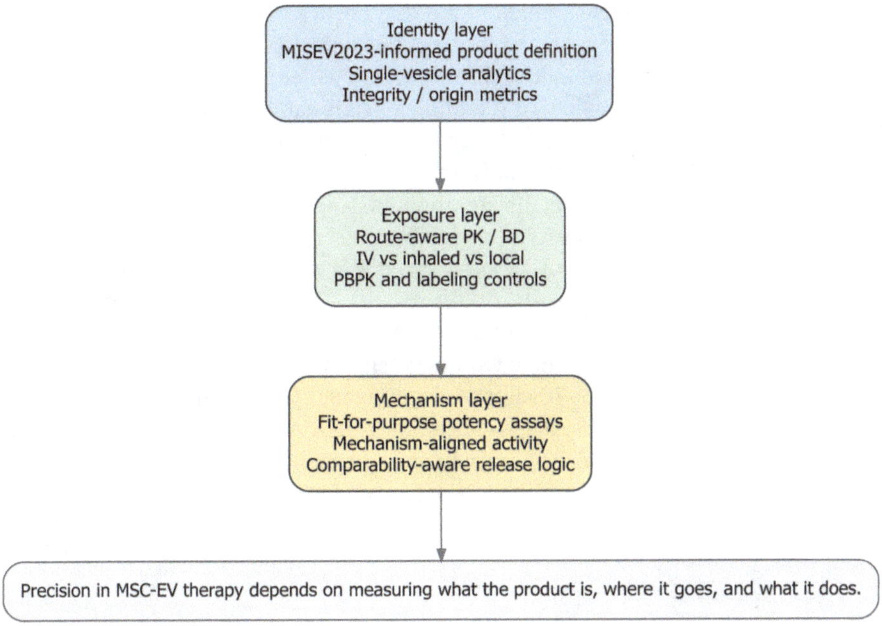

Fig. 12.2 Precision MSC–EV therapeutics as a three-layer system: identity, exposure, and mechanism. This figure conceptualizes precision MSC–EV therapy as the integration of three linked layers: product identity, route-aware exposure, and mechanism-aligned activity. Updated minimal information standards, single-vesicle analytics, and integrity metrics define what the product is; pharmacokinetic and biodistribution frameworks determine where it goes; and fit-for-purpose potency assays determine what it does. The figure emphasizes that "precision" in MSC–EV therapy is not limited to disease targeting, but depends equally on measurement truthfulness, exposure definition, and mechanistic alignment

A third challenge is potency definition for pleiotropic MoA. The field widely recognizes that MSC–EV effects can be multi-component and context specific, making single "one-size potency assays" risky. Guidance emphasizes fit-for-purpose potency tests aligned to plausible mechanisms and intended indications [131].

Regulatory and ethical considerations. From a regulatory perspective, precision hinges on CQAs, comparability, and evidence packages that resemble biologics development: EV-focused medicinal product guidance outlines manufacturing, analytics, nonclinical, and early clinical trial considerations aligned with European legislation (EVOLVE France) [107]. In addition, a major open-access assessment from the Pharmaceuticals and Medical Devices Agency science board clarifies key qual ity/safety gaps and proposes a stepwise roadmap spanning manufacturing, characterization, nonclinical, and clinical evaluation [23, 229].

Ethically, "precision therapy" increases the likelihood that EV products become coupled to high-dimensional donor/patient data (omics, imaging, AI-derived stratification), elevating privacy and consent stakes even when the therapy itself is cell-free. The exRNA program landscape analysis highlights the translational trajectory of biofluid exRNA/EV measurements into diagnostics and products, underscoring the need for governance as data and therapeutics converge [290].

Translational and clinical evidence. Precision challenges are visible in trial heterogeneity. A systematic review of EV clinical trials catalogued 471 EV-related trials across >200 diseases, noting that diagnostics dominate but that therapeutic trials frequently use MSC–EVs, commonly in respiratory indications—and also highlighting that methodological details are often missing from trial records [291]. This aligns with the need for standardized reporting and precision-grade metadata (EV-TRACK; MIBlood-EV for blood-derived EV research as a methodological comparator) [106].

Future research directions. Short-term (1–3 years): establish consensus therapeutic dose metrics and cross-validated single-vesicle QC panels for release/bridging studies [159]. Mid-term (3–7 years): develop and validate mechanism-aligned potency assay matrices usable across sites, integrated with comparability protocols [131]. Long-term (7–15 years): formalize EV-specific reference materials and regulatory-grade standards enabling true inter-lab comparability (counts, refractive index, epitope abundance), and link exposure–response quantitatively using PBPK + clinical biomarkers ("EV pharmacology as biologic pharmacology") [292].

12.2 Engineering and Targeting

State of the art. MSC–EVs are now widely treated as programmable nanobiologics: their cargo and surfaces can be modified before isolation (engineering donor MSCs) or after isolation (post-production loading/functionalization). Engineering goals typically fall into four classes: (i) increase yield and manufacturing robustness, (ii) increase cargo load and functional delivery, (iii) increase tissue/cell targeting and reduce clearance, and (iv) control spatiotemporal release [293] (Table 12.1).

Key recent advances. A landmark manufacturing-and-engineering advance is cellular nanoporation enabling large-scale generation of functional mRNA-encapsulating EVs [294], addressing a core limitation of post-isolation loading inefficiency for large RNAs and improving scalability.

A second inflection point is clinical validation of engineered exosome therapeutics. A first-in-human phase I study reported engineered EVs loaded with KRAS^G12D^ siRNA [286], with corresponding trial registration (NCT03608631), demonstrating feasibility of engineered EVs as a precision nucleic acid delivery biologic.

Third, modular CRISPR/Cas delivery approaches using EV systems have rapidly matured, including modular strategies for Cas9 loading and delivery [295] and broader reviews that organize loading, surface functionalization, and delivery barriers for genome editing.

Table 12.1 Comparison of engineered EV strategies relevant to MSC–EV therapeutics

Engineered EV strategy (stage)	Typical cargo types	Targeting ligands/ targeting concept	Advantages	Key limitations/risks	Key recent refs.
Producer cell genetic cargo programming (pre-isolation)	miRNA/siRNA, mRNA, proteins	Indirect (use native tropism or co-engineer surface)	Higher cargo encapsulation consistency vs post-loading; can be scaled with stable cell lines	Risk of altering MSC biology and EV composition broadly; comparability when cell line drifts; regulatory scrutiny of genetic modification	[293, 294]
Cellular nanoporation for mRNA EVs (pre-isolation)	Therapeutic mRNA; co-expressed targeting peptides	Genetically encoded targeting peptides	Strong improvement in mRNA payload levels and EV yields; manufacturing-relevant approach	Requires specialized equipment/ process control; still needs robust release and potency assays	[294]
Clinical engineered siRNA EVs (pre-isolation + loading)	KRAS^G12D^ siRNA	Exploit pancreas biodistribution; engineered exosome platform	Demonstrates clinical feasibility of engineered EV nucleic acid medicine	Disease-specific targeting; scale-up, long-term safety, and potency definition still needed	[286] NCT03608631
"Don't-eat-me" shielding to reduce MPS clearance (pre-isolation)	Native + co-loaded RNAs/miRNAs	CD47 overexpression concept	Potentially increases circulation time and delivery efficiency by reducing macrophage uptake	Immunological implications; needs careful safety evaluation; mechanism may vary by route/ dose	[105, 298]
Post-isolation electroporation/physical loading	siRNA/miRNA, small molecules	Usually none unless combined with surface targeting	Simple, adaptable to many cargos	Can damage EV membranes; aggregation; variable loading efficiency; needs rigorous cargo quantification	[299]
Post-isolation membrane fusion/hybrid EV–nanoparticle formulations	mRNA, siRNA, proteins	Hybridization can add synthetic targeting modules	Higher nucleic acid loading and delivery potential; bridges EV biocompatibility with synthetic efficiency	Product definition is harder (hybrid identity); regulatory classification and comparability are complex	[300]

(continued)

Table 12.1 (continued)

Engineered EV strategy (stage)	Typical cargo types	Targeting ligands/ targeting concept	Advantages	Key limitations/risks	Key recent refs.
Genetic surface display via EV-enriched proteins (pre-isolation)	Any (payload independent)	Fusions to Lamp2b/ CD63/CD9, etc. presenting peptides/ ligands	Stable ligand display; enables "built-in" targeting	Can change EV biology; needs single-vesicle verification of display density	[293, 301]
Click chemistry surface functionalization (post-isolation or metabolic tagging)	RNA/drugs/ proteins; imaging agents	Conjugated peptides/ antibodies/aptamers; bio-orthogonal handles	High modularity; supports multiplex targeting and tracking	Copper toxicity concerns; residual reagents; standardization of conjugation stoichiometry	[302]
Targeted capture/ purification of specific EV phenotypes (downstream engineering/QC)	N/A (process)	Immunoaffinity capture	Increases product consistency; may enrich active subpopulation	Scalability and cost; risk of selecting incomplete biology; defines product by marker	[126]
Controlled release via hydrogels/biomaterials (formulation level)	Native or engineered EVs	Local retention rather than molecular ligand targeting	Extends residence time; reduces dosing frequency; can enable local tissue repair	Requires material–EV compatibility validation; complex CMC (combination product questions)	[297]
Scale-compatible engineering + isolation workflows (process integration)	DNA/RNA payloading, surface displays	Compatible with TFF-based downstream	Moves engineered EVs closer to manufacturable reality	TFF shear/retention and recovery optimization; analytics needed to show no function loss	[303]

Fourth, the engineering toolbox has become more chemistry enabled, especially via click chemistry surface modification. New methodological work emphasizes both power and risk, while therapeutic demonstrations include click-engineered sEVs expressing CARs with enhanced efficacy in acute liver failure models [296].

Fifth, “EVs as formulations” has matured through delivery matrices: hydrogel-based sustained release and metabolic tagging approaches provide controllable retention and release kinetics in vivo, addressing the frequent mismatch between EV clearance and desired therapeutic duration [297].

Technical challenges. Engineering raises the bar for analytics and comparability. As EV products become more complex (surface ligands + nucleic acid payloads + formulation matrices), it becomes harder to prove which attributes drive efficacy and to keep those attributes stable across scale-up and manufacturing changes—precisely the CMC problem highlighted in medicinal product development guidance [107].

A second challenge is targeting realism. Biodistribution reviews emphasize that EVs often accumulate in clearance organs and face biological barriers; engineering must overcome both uptake and cytosolic delivery limitations, which are not guaranteed by receptor binding alone [270].

A third challenge is immunogenicity and immune recognition. Although EVs are often described as low immunogenic, immunogenicity depends on origin, surface molecules, and engineering choices; strategies to reduce recognition can conflict with potency or safety goals [304].

Regulatory and ethical considerations. Engineering choices create regulatory implications: genetic modification and chemical conjugation can introduce new impurities, residual reagents, or altered biological activities that must be controlled. EV-focused quality/safety assessments stress the need for systematic evaluation of contaminants, degradation, and heterogeneity—especially when single-particle distributions, not just averages, matter [23].

Ethically, engineering that meaningfully changes biodistribution or carries gene editing cargo must be accompanied by careful consideration of off-target exposure and long-term surveillance principles aligned with advanced therapy development norms [299].

Translational and clinical evidence. Beyond the KRAS^G12D^ siRNA engineered exosome phase I experience ([286]; NCT03608631), the broader clinical landscape shows growth but uneven reporting. Trial landscape analyses emphasize that many therapeutic trials rely on MSC–EVs, yet records often lack detailed isolation/characterization metadata, limiting cross-trial inference and highlighting the need for better engineered product descriptors [291].

Future research directions. Short-term: establish standardized reporting and quantification of engineering degree (ligand density distributions, cargo copies per EV, fraction of EVs successfully modified), anchored to single-vesicle measurement [288]. Mid-term: develop engineering-aware potency assays that measure functional delivery (endosomal escape, cytosolic activity) rather than uptake; incorporate quantitative delivery assays already emerging in EV biology [289]. Long-term: converge on “modular EV platforms” with predictable behavior (platform

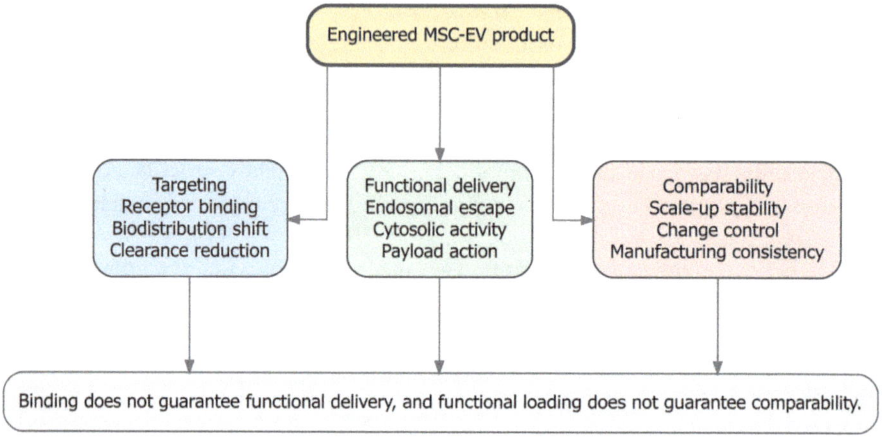

Fig. 12.3 The engineered MSC–EV control problem: targeting, delivery, and comparability are not the same challenge. This figure separates three often conflated engineering goals in MSC–EV development: tissue targeting, intracellular functional delivery, and manufacturing comparability. Surface binding or receptor engagement does not guarantee endosomal escape or cytosolic activity, and successful functional loading does not automatically guarantee reproducible scale-up. By distinguishing these layers, the figure clarifies why engineered EV programs require engineering-aware potency assays and comparability strategies rather than relying only on uptake or biodistribution data

master file concept)—stable producer lines + defined engineering modules + validated comparability, enabling rapid indication expansion without re-inventing CMC for each target [107] (Fig. 12.3).

12.3 AI and Omics Integration

State of the art. Multi-omics and AI are increasingly central to MSC–EV therapy for three roles: (i) mechanism and active component discovery (proteomics/transcriptomics/lipidomics/metabolomics), (ii) precision stratification and companion diagnostics (EV signatures predicting who responds), and (iii) manufacturing control (high-dimensional analytics for batch release and drift detection). This convergence mirrors the broader translational trajectory of exRNA/EV technologies into diagnostics and products described by the NIH exRNA program's decade analysis [290].

Key infrastructure enabling AI/omics integration includes curated EV cargo databases: Vesiclepedia (updated to include EVs and additional extracellular particles) and ExoCarta (recently updated with expanded entries and quantitative plugins), alongside exRNA repositories (exRNA Atlas) that standardize processing pipelines and metrics [305].

12.3.1 Key Recent Advances

First, database maturity has improved markedly. Vesiclepedia 2024 reports expansion to thousands of EV studies and hundreds of thousands of molecular entries, and ExoCarta's update adds quantitative comparison tools, improving feasibility of systematic "omics → candidate attribute" workflows [305].

Second, computational reviews now frame multi-omics integration strategies (rule-based filtering through deep learning) for EV biomarker selection and, by extension, for therapy design decisions such as target selection and tissue specificity mapping [306].

Third, machine learning is becoming a practical tool for high-dimensional EV measurement modalities, including imaging and omics data; newer reviews explicitly address ML's role not only in diagnostics but also in manufacturing QC and therapeutic development [307].

Fourth, emerging studies and reviews emphasize the importance of single-EV analytics + computation, where multiplexed, high-throughput single-vesicle datasets demand dimensionality reduction, clustering, and robust calibration workflows [288].

Technical challenges. A major challenge is data comparability across platforms and laboratories. EV datasets are susceptible to batch effects from pre-analytics (collection, processing), separation methods, and measurement platforms; minimal-information reporting standards (MISEV2023; MIBlood-EV) and EV-TRACK aim to mitigate this, but adoption is inconsistent [71].

A second challenge is causality vs correlation: omics signatures frequently correlate with outcomes but may not drive potency. The therapeutic field's experience—where active vs inactive MSC–EV lots can differ in vivo without clear surrogate markers—illustrates why AI must be integrated with mechanistic assays and not used as a stand-alone gatekeeper [287].

A third challenge is that "AI-ready endpoints" are not yet standardized for EV therapeutics. Without consistent dose metrics, functional delivery readouts, and potency assays, supervised learning can learn artifacts rather than biology [131].

Tamadon et al. [308] outlined the expanding role of artificial intelligence in regenerative medicine, including image analysis, multi-omics integration, predictive modeling, and treatment stratification. In MSC–EV development, these tools are most credible when applied to standardized datasets that connect producer cell characteristics, manufacturing conditions, EV molecular profiles, potency measurements, biodistribution, and clinical outcomes.

12.3.2 Regulatory and Ethical Considerations

As AI enters product development, regulators typically require transparency, validation, and control of model drift—especially when models influence release decisions or patient stratification. While EV-specific AI regulatory guidance remains emergent, EV medicinal product and quality/safety assessments underline the necessity of traceability and reproducibility, which AI can either strengthen (automated QC) or weaken (opaque models) depending on governance [107]. Ethically, AI/omics integration raises issues of privacy, consent for secondary use of molecular data, and fairness in model performance across populations, particularly if EV-based companion diagnostics influence access to regenerative therapies. The exRNA program experience—moving exRNA technologies toward clinical and commercial use—highlights the speed at which these concerns become practical [290].

12.3.3 Translational and Clinical Evidence

Although most EV clinical trials historically focus on diagnostics, the EV trial landscape review documents rapid growth and shows that therapeutic trials often employ MSC–EVs, implying a near-term need for standardized datasets that can support AI-enabled meta-analyses and cross-trial learning [291].

12.3.4 Future Research Directions

Short-term: create standardized, shareable "EV therapeutic development datasets" pairing multi-omics with validated potency assays and single-vesicle QC outputs, in MISEV/EV-TRACK-aligned formats [71]. Mid-term: build "design–build–test–learn" cycles for engineered MSC–EVs—AI proposes cargo/targeting modules → manufacturing generates variants → standardized assays measure potency/delivery → models update [294]. Long-term: develop regulatory-acceptable "digital comparability" tools that detect drift in EV products using multivariate signatures (single-vesicle phenotype distributions + multi-omics + functional assays), enabling platform-based approvals and faster indication expansion [23] (Fig. 12.4).

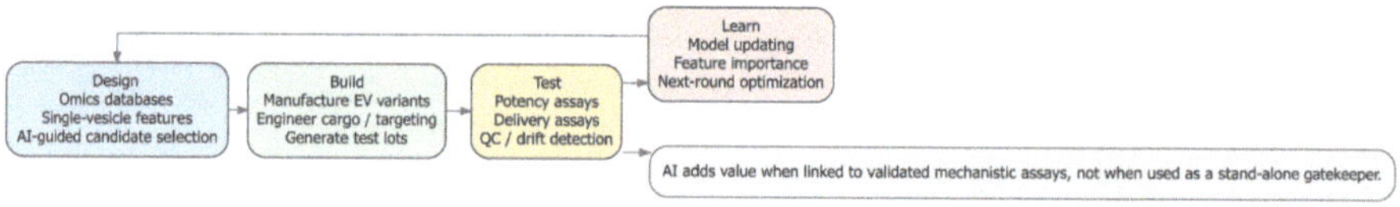

Fig. 12.4 AI-enabled design–build–test–learn cycle for MSC–EV therapeutic development. This figure presents an AI-enabled design–build–test–learn workflow for MSC–EV development. Multi-omics databases, single-vesicle analytics, and curated EV repositories support candidate discovery; manufacturing generates variant products; functional and delivery assays provide mechanistic outputs; and machine learning updates design rules for subsequent optimization cycles. The figure emphasizes that AI is most useful when integrated with validated potency and delivery assays, rather than treated as a stand-alone predictive layer

12.4 Personalized Regenerative Medicine

State of the art. In regenerative medicine, MSC–EVs are often positioned as off-the-shelf, lower-risk alternatives to viable cell administration (reduced concerns about ectopic engraftment and some cell therapy risks). However, their clinical translation remains constrained by heterogeneity in production, characterization, dosing, and trial designs [309]. Personalization in MSC–EV therapy is now trending toward three pragmatic meanings: (1) personalized by indication and route (matching dosing metrics and formulations to tissue exposure), (2) personalized by biological state (priming/conditioning MSCs to produce EVs aligned with desired mechanisms), and (3) personalized by patient stratification (biomarkers predicting who benefits or who requires modified dosing) [159].

12.4.1 Key Recent Advances

"Personalization by route" has advanced through inhalation/nebulization clinical programs. Preclinical and early clinical work demonstrated feasibility and short-term safety of nebulized allogeneic adipose MSC–EVs in healthy volunteers with supporting lung injury models ([163]; NCT04313647). More recently, a clinical investigation of nebulized umbilical cord MSC–EVs reported lung accumulation in animal models and symptomatic/CT improvements in pulmonary fibrosis patients [165].

"Personalization by patient stratification" is still early, but randomized placebo-controlled evidence is emerging in defined populations, such as EV infusion (ExoFlo) in COVID-19 respiratory failure (NCT04493242) [161].

"Personalization by biological state" has advanced with priming paradigms. Studies show that inflammatory conditioning can yield MSC–EVs with superior immunomodulatory function compared to single cytokine priming, including in transplant models, and can identify mechanistic miRNA contributors whose knockdown diminishes efficacy [310].

At the evidence synthesis level, a 2026 systematic review and meta-analysis of human MSC–EV trials (24 interventional trials; $n \approx 536$) estimated low serious adverse event rates and low adverse event rates while emphasizing that heterogeneous manufacturing and trial designs limit definitive conclusions [229].

12.4.2 Technical Challenges

The dominant barrier remains lot variability and lack of validated potency assays. Standardized production does not guarantee consistent function: independent MSC–EV preparations can differ in immunomodulatory activity in vitro and in vivo, and attempts to find simple surrogate markers may fail [287]. This directly motivates assay matrices and mechanism-aligned potency tests, as emphasized by ISCT-linked guidance [131].

A second challenge is identity potency decoupling. For example, CD73 activity in MSC–EV preparations may be detergent resistant yet not correlate with immunomodulatory capabilities, illustrating why identity markers alone cannot be treated as potency proxies [311].

A third challenge is immunogenicity and clearance, especially for repeated dosing or systemic routes. EV immunogenicity depends on cellular origin and surface composition; strategies to reduce clearance (including CD47-related approaches) must be balanced against immune risk and long-term surveillance [304].

A fourth challenge is stability and distribution. Systematic reviews show that storage and freeze–thaw protocols can alter EV concentration, aggregation, cargo, and bioactivity; lyophilization strategies are being explored to reduce cold chain dependence and improve point-of-care feasibility—critical for personalized regenerative applications that may require decentralized delivery [153].

12.4.3 Regulatory and Ethical Considerations

Personalization increases complexity for CMC and oversight. EV medicinal product guidance emphasizes dossier requirements and early-phase clinical development needs; Japan's PMDA science board assessment frames a stepwise roadmap and highlights heterogeneity and contamination risks, recommending careful upstream control and avoidance of animal-derived materials where possible [107].

Ethically, personalization raises questions of equity and access (especially if complex analytics and cold chain–dependent products become cost drivers) and data governance when omics-based stratification and AI are used to match patients to products [290].

Translational and clinical evidence. Clinical registries show a growing number of MSC–EV trials across routes and indications (examples include ocular delivery NCT04213248; stroke NCT03384433; inhalation studies NCT04276987; nebulization safety NCT04313647; systemic infusion NCT04493242). Evidence synthesis indicates generally favorable short-term safety signals but highlights inadequate standardization and underpowered designs as key limitations [229].

12.4.4 Future Research Directions

Short-term: define "personalization primitives" usable across indications—validated priming recipes (inputs → EV phenotype), minimal potency assay panels per MoA class, and stability-preserving formulations enabling clinic-friendly delivery [310]. Mid-term: run adaptive, biomarker-informed trials where EV dosing and

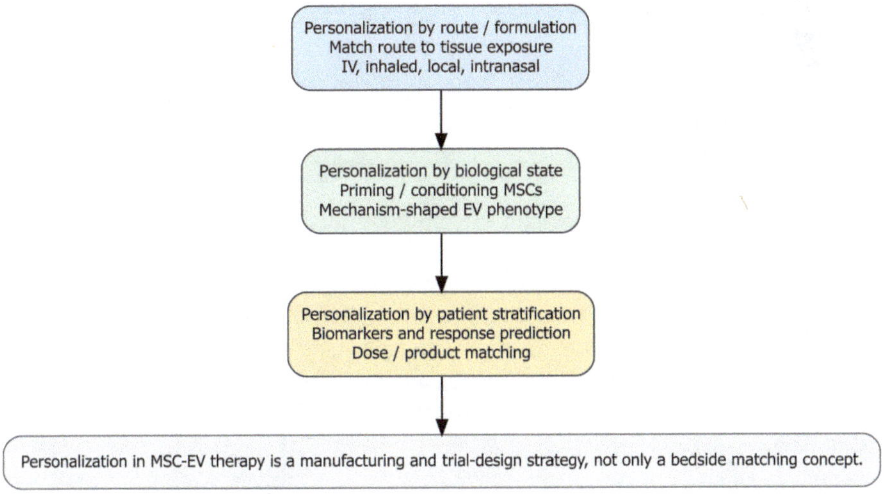

Fig. 12.5 Three practical forms of personalization in MSC–EV regenerative therapy. This figure organizes personalization in MSC–EV therapy into three pragmatic domains: personalization by route and formulation, personalization by biological state of the producer cell, and personalization by patient stratification. Route/formulation matching addresses tissue exposure; priming and conditioning adjust the biological state of the EV-producing MSCs; and biomarker-based stratification determines who may benefit and under what dosing logic. The figure reframes personalization as a manufacturing and trial design strategy, not only a bedside matching concept

priming are adjusted based on early response markers; develop exposure–response models per route using validated biodistribution methods [270]. Long-term: establish platform-based MSC–EV therapies—standardized donor/producer systems with modular priming and engineering, supported by multi-site comparability, enabling precision regenerative medicine at scale [107].

These three practical forms of personalization in MSC–EV regenerative therapy are summarized in Fig. 12.5.

12.5 Future Clinical Adoption Roadmap

State of the art. The field is approaching a transition from exploratory proof-of-concept to disciplined product development. EV clinical trial landscape data show rapid growth and substantial heterogeneity; parallel guidance documents and society-driven initiatives increasingly emphasize early regulatory engagement, mechanism clarity, potency integration into manufacturing, and market viability [291].

12.5.1 Key Recent Advances

Clinical adoption is aided by the emergence of clinical-grade manufacturing and QC studies, including GLP-grade assessments and systematic preclinical safety work for MSC–EV products [312]. It is also supported by practical trial outcomes: nebulized MSC–EV safety data in healthy volunteers (Shi et al., 2021; NCT04313647) and randomized controlled data for EV infusion in COVID-19 respiratory failure (NCT04493242) [161, 163].

At the system level, EV medicinal product development recommendations (EVOLVE France) and PMDA's scientific assessment outline concrete expectations for manufacturing, QC, and evaluation strategies—effectively translating EV science into "CMC + nonclinical + clinical" language [107].

Technical challenges. The roadmap's technical friction points are predictable but unresolved:

- Scalable, closed, GMP-compatible manufacturing with consistent recovery and minimized contaminants [313].
- Validated potency assays and reference materials that support batch release and comparability through change (bioreactor scale-up, donor changes, engineering updates) [131].
- Stability and logistics solutions that reduce cold chain dependence and preserve function, enabling broader access (lyophilization/optimized storage) [153].
- Robust nonclinical safety packages tailored to route and risks (immunogenicity, thrombogenicity, off-target biodistribution) [304].

12.5.2 Regulatory and Ethical Considerations

A practical adoption roadmap must decide early whether MSC–EV products are framed as biologic-like therapeutics, drug delivery platforms, or combination products (when combined with hydrogels or devices), because this determines documentation and evaluation expectations. EVOLVE France and PMDA-aligned assessments emphasize early-phase dossier contents, analytical strategies, and quality frameworks consistent with medicinal product development [107].

Ethically, adoption requires mitigation of "therapy tourism" and unsubstantiated clinical offerings by anchoring clinical claims to registered trials and peer-reviewed evidence. Trial landscape studies show rapid growth and variability, which can be exploited without strong standards and enforcement [291].

Fig. 12.6 Clinical adoption roadmap for MSC–EV therapy from exploratory biologics to platform regenerative medicine. This roadmap figure shows the staged path from exploratory MSC–EV products to clinically adopted platform therapies. Early stages focus on reporting compliance, calibrated quantification, stability planning, and route-specific safety packages; intermediate stages emphasize multicenter ring trials, potency harmonization, and comparability through manufacturing change; long-term stages move toward platform approvals with standardized producer systems, modular engineering, and validated cross-indication expansion. The figure highlights that future adoption depends on aligning quality, analytics, engineering, and regulation into one disciplined biologics-like development pathway

12.5.3 Translational and Clinical Evidence

A disease-agnostic but illustrative evidence base now spans multiple routes and indications, with registry-backed trials and published outcomes. A 2026 meta-analysis across MSC–EV human trials suggests low pooled SAE and AE rates, while emphasizing that improved standardization and larger RCTs are needed for definitive efficacy [229].

Future research directions. Short-term: implement a "clinical-readiness checklist" for MSC–EV products—reporting compliance, calibrated quantification, stability plan, potency assay justification, and route-specific safety assays [71]. Mid-term: execute multicenter ring trials for potency and counting methods (interlaboratory calibration + reference materials), and embed these into clinical trial CMC sections [314]. Long-term: develop platform approvals—standardized MSC producer systems (or derived stable lines), modular targeting/payloading, and validated comparability, enabling rapid translation across indications while maintaining safety and efficacy [107].

The staged pathway from exploratory MSC–EV biologics to platform-based regenerative medicine is summarized in Fig. 12.6.

12.6 Conclusion

MSC–EV-based cell-free therapy is entering a decisive maturation phase in which the main obstacle is no longer whether EVs can modulate biology, but whether the field can define, manufacture, measure, and regulate EV products with the discipline expected of modern biologics. Over the last decade, foundational standards (MISEV2018 and the expanded MISEV2023) and transparency initiatives (EV-TRACK; blood–EV reporting tools) have clarified what must be reported and measured to make claims reproducible. Simultaneously, clinical experience has become tangible: nebulized clinical-grade MSC–EVs have demonstrated short-term tolerability in healthy volunteers and supportive efficacy in animal lung injury models; randomized controlled EV infusion has produced safety and signal-of-benefit data in COVID-19 respiratory failure; and engineered EVs delivering KRAS-targeting siRNA have reached phase I human testing, signaling feasibility of precision, payloaded EV medicines. Yet these advances also expose the central translational constraint: identity is not potency—functionally distinct MSC–EV lots can arise under ostensibly standardized processes, and simple surrogate markers may not track immunomodulatory efficacy.

The most credible future pathway therefore integrates (i) single-vesicle and calibrated analytics to control heterogeneity, (ii) mechanism-aligned potency assay matrices embedded early in process development and maintained through comparability, (iii) engineering approaches quantified at the single-particle level, and (iv) delivery/stability solutions (including lyophilization and controlled-release formats) that make EV therapies accessible beyond specialized centers. Achieving this would reposition MSC–EVs from promising acellular biologics to a platform regenerative modality with predictable performance, defensible regulation, and scalable clinical impact.

12.6.1 Open Questions

What are the minimal CQAs that predict clinical activity across indications, and which require indication-specific tailoring [107]?

Which potency assay panels best balance mechanistic relevance, robustness, and regulatory acceptability for pleiotropic MSC–EV products [131]?

How should "dose" be standardized across routes and platforms to enable exposure–response modeling and meta-analysis [159]?

What single-vesicle metrics (marker distributions, cargo distributions, functional delivery proxies) should become routine release tests as instrumentation matures [288]?

How can engineered EV products maintain comparability through iterative modifications (targeting ligands, payloads, formulations) without resetting the full clinical development pathway [107]?

References

1. Wu X, Jiang J, Gu Z, Zhang J, Chen Y, Liu X. Mesenchymal stromal cell therapies: immunomodulatory properties and clinical progress. Stem Cell Res Ther. 2020;11(1):345. https://doi.org/10.1186/s13287-020-01855-9.
2. Kirkeby A, Main H, Carpenter M. Pluripotent stem-cell-derived therapies in clinical trial: a 2025 update. Cell Stem Cell. 2025;32(1):10–37. https://doi.org/10.1016/j.stem.2024.12.005.
3. U.S. Food and Drug Administration. Regulatory considerations for human cells, tissues, and cellular and tissue-based products: minimal manipulation and homologous use guidance for industry and Food and Drug Administration staff. Silver Spring, MD: U.S. Department of Health and Human Services; 2017.
4. U.S. Food and Drug Administration. FDA approves remestemcel-L-rknd for steroid-refractory acute graft versus host disease in pediatric patients. U.S. Food and Drug Administration; 2024.
5. Thery C, Witwer KW, Aikawa E, Alcaraz MJ, Anderson JD, Andriantsitohaina R, et al. Minimal information for studies of extracellular vesicles 2018 (MISEV2018): a position statement of the International Society for Extracellular Vesicles and update of the MISEV2014 guidelines. J Extracell Vesicles. 2018;7(1):1535750. https://doi.org/10.1080/20013078.2018.1535750.
6. Singh VK, Saini A, Kalsan M, Kumar N, Chandra R. Describing the stem cell potency: the various methods of functional assessment and in silico diagnostics. Front Cell Dev Biol. 2016;4:134. https://doi.org/10.3389/fcell.2016.00134.
7. Viswanathan S, Shi Y, Galipeau J, Krampera M, Leblanc K, Martin I, et al. Mesenchymal stem versus stromal cells: International Society for Cell & Gene Therapy (ISCT(R)) Mesenchymal Stromal Cell committee position statement on nomenclature. Cytotherapy. 2019;21(10):1019–24. https://doi.org/10.1016/j.jcyt.2019.08.002.
8. Dhaiban S, Chandran S, Noshi M, Sajini AA. Clinical translation of human iPSC technologies: advances, safety concerns, and future directions. Front Cell Dev Biol. 2025;13:1627149. https://doi.org/10.3389/fcell.2025.1627149.
9. Aljagthmi AA, Abdel-Aziz AK. Hematopoietic stem cells: understanding the mechanisms to unleash the therapeutic potential of hematopoietic stem cell transplantation. Stem Cell Res Ther. 2025;16(1):60. https://doi.org/10.1186/s13287-024-04126-z.
10. Sawamoto N, Doi D, Nakanishi E, Sawamura M, Kikuchi T, Yamakado H, et al. Phase I/II trial of iPS-cell-derived dopaminergic cells for Parkinson's disease. Nature. 2025;641(8064):971–7. https://doi.org/10.1038/s41586-025-08700-0.
11. Weiss ARR, Dahlke MH. Immunomodulation by Mesenchymal Stem Cells (MSCs): mechanisms of action of living, apoptotic, and dead MSCs. Front Immunol. 2019;10:1191. https://doi.org/10.3389/fimmu.2019.01191.

N. M. Mussin, A. Tamadon, *Mesenchymal Stem Cells and Their Extracellular Vesicles*, https://doi.org/10.1007/978-3-032-36437-1

12. Guan Q, Li Y, Shpiruk T, Bhagwat S, Wall DA. Inducible indoleamine 2,3-dioxygenase 1 and programmed death ligand 1 expression as the potency marker for mesenchymal stromal cells. Cytotherapy. 2018;20(5):639–49. https://doi.org/10.1016/j.jcyt.2018.02.003.
13. Hernandez Pichardo A, Wilm B, Liptrott NJ, Murray P. Intravenous administration of human umbilical cord mesenchymal stromal cells leads to an inflammatory response in the lung. Stem Cells Int. 2023;2023:7397819. https://doi.org/10.1155/2023/7397819.
14. Cheung TS, Giacomini C, Cereda M, Avivar-Valderas A, Capece D, Bertolino GM, et al. Apoptosis in mesenchymal stromal cells activates an immunosuppressive secretome predicting clinical response in Crohn's disease. Mol Ther. 2023;31(12):3531–44. https://doi.org/10.1016/j.ymthe.2023.10.004.
15. U.S. Food and Drug Administration, Marrero-Berrios I. Summary basis for regulatory action: remestemcel-L-rknd (RYONCIL) for steroid-refractory acute graft versus host disease in pediatric patients. U.S. Food and Drug Administration, Office of Therapeutic Products; 2024.
16. Guillamat-Prats R. Role of mesenchymal stem/stromal cells in coagulation. Int J Mol Sci. 2022;23(18) https://doi.org/10.3390/ijms231810393.
17. Foti R, Storti G, Palmesano M, Calicchia A, Foti R, Ciprandi G, et al. AI-driven innovations for quality control and standardization: future strategies in adipose-derived stem cell manufacturing. Int J Mol Sci. 2026;27(5) https://doi.org/10.3390/ijms27052388.
18. European Commission. EudraLex volume 4: guidelines on good manufacturing practice specific to advanced therapy medicinal products. European Commission; 2017.
19. European Medicines Agency. Alofisel withdrawn from the EU market. European Medicines Agency; 2024.
20. Takeda Pharmaceutical Company Limited. Takeda provides an update on Alofisel (darvadstrocel); 2024.
21. International Society for Extracellular Vesicles. Minimal information for studies of extracellular vesicles (MISEV2023): from basic to advanced approaches. International Society for Extracellular Vesicles; 2023.
22. Upadhya D, Shetty AK. MISEV2023 provides an updated and key reference for researchers studying the basic biology and applications of extracellular vesicles. Stem Cells Transl Med. 2024;13(9):848–50. https://doi.org/10.1093/stcltm/szae052.
23. Takakura Y, Hanayama R, Akiyoshi K, Futaki S, Hida K, Ichiki T, et al. Quality and safety considerations for therapeutic products based on extracellular vesicles. Pharm Res. 2024;41(8):1573–94. https://doi.org/10.1007/s11095-024-03757-4.
24. Lopez-Garcia L, Castro-Manrreza ME. TNF-alpha and IFN-gamma participate in improving the immunoregulatory capacity of mesenchymal stem/stromal cells: importance of cell-cell contact and extracellular vesicles. Int J Mol Sci. 2021;22(17) https://doi.org/10.3390/ijms22179531.
25. U.S. Food and Drug Administration. Potency assurance for cellular and gene therapy products: draft guidance for industry. U.S. Department of Health and Human Services; 2023.
26. U.S. Food and Drug Administration. Manufacturing changes and comparability for human cellular and gene therapy products: draft guidance for industry. U.S. Department of Health and Human Services; 2023.
27. U.S. Food and Drug Administration. Same surgical procedure exception under 21 CFR 1271.15(b): questions and answers regarding the scope of the exception: guidance for industry. U.S. Department of Health and Human Services; 2020.
28. 21st Century Cures Act (Public Law 114–255); 2016.
29. U.S. Food and Drug Administration. Regenerative medicine advanced therapy (RMAT) designation. U.S. Food and Drug Administration; 2025.
30. U.S. Food and Drug Administration. Framework for the regulation of regenerative medicine products. U.S. Food and Drug Administration; 2019.
31. European Commission. Good manufacturing practice for advanced therapy medicinal products (EudraLex Volume 4). Brussels: European Commission; 2017.

32. European Medicines Agency. Guideline on quality, non-clinical and clinical requirements for investigational advanced therapy medicinal products in clinical trials. Amsterdam: European Medicines Agency; 2025.
33. Han X, Liao R, Li X, Zhang C, Huo S, Qin L, et al. Mesenchymal stem cells in treating human diseases: molecular mechanisms and clinical studies. Signal Transduct Target Ther. 2025;10(1):262. https://doi.org/10.1038/s41392-025-02313-9.
34. Direct Biologics LLC. A phase I study of ExoFlo, an ex vivo culture-expanded adult allogeneic bone marrow mesenchymal stem cell derived extracellular vesicle isolate product, for the treatment of medically refractory ulcerative colitis: ClinicalTrials.gov. U.S. National Library of Medicine; 2025 [updated 2025/02/11]. Available from: https://clinicaltrials.gov/study/NCT05176366.
35. Huazhong University of Science and Technology. Clinical study of the safety and efficacy of umbilical cord mesenchymal stem cell exosomes in treating chronic cough after COVID-19 infection; 2023.
36. Mussin NM, Zhilisbayeva KR, Baspakova A, Kurmanalina MA, Tamadon A. Mapping the landscape of mesenchymal stem cell-derived extracellular vesicles: from bench to bedside. Stem Cells Int. 2025;2025:5361754. https://doi.org/10.1155/sci/5361754.
37. Herbst L, Niessing B, Schmitt RH. Identification of critical process parameters and quality attributes for bioreactor-based expansion of human MSCs. Front Bioeng Biotechnol. 2025;13:1608194. https://doi.org/10.3389/fbioe.2025.1608194.
38. Olmedo-Moreno L, Aguilera Y, Balina-Sanchez C, Martin-Montalvo A, Capilla-Gonzalez V. Heterogeneity of in vitro expanded mesenchymal stromal cells and strategies to improve their therapeutic actions. Pharmaceutics. 2022;14(5) https://doi.org/10.3390/pharmaceutics14051112.
39. Petrenko Y, Vackova I, Kekulova K, Chudickova M, Koci Z, Turnovcova K, et al. A comparative analysis of multipotent mesenchymal stromal cells derived from different sources, with a focus on neuroregenerative potential. Sci Rep. 2020;10(1):4290. https://doi.org/10.1038/s41598-020-61167-z.
40. Galleu A, Riffo-Vasquez Y, Trento C, Lomas C, Dolcetti L, Cheung TS, et al. Apoptosis in mesenchymal stromal cells induces in vivo recipient-mediated immunomodulation. Sci Transl Med. 2017;9(416):eaam7828. https://doi.org/10.1126/scitranslmed.aam7828.
41. Shan Y, Zhang M, Tao E, Wang J, Wei N, Lu Y, et al. Pharmacokinetic characteristics of mesenchymal stem cells in translational challenges. Signal Transduct Target Ther. 2024;9(1):242. https://doi.org/10.1038/s41392-024-01936-8.
42. Thompson M, Mei SHJ, Wolfe D, Champagne J, Fergusson D, Stewart DJ, et al. Cell therapy with intravascular administration of mesenchymal stromal cells continues to appear safe: an updated systematic review and meta-analysis. EClinicalMedicine. 2020;19:100249. https://doi.org/10.1016/j.eclinm.2019.100249.
43. Caplan AI. Mesenchymal stem cells: time to change the name! Stem Cells Transl Med. 2017;6(6):1445–51. https://doi.org/10.1002/sctm.17-0051.
44. Phinney DG, Hwa Lee R, Boregowda SV. Revisiting the mesenchymal "stem vs. stromal" cell dichotomy and its implications for development of improved potency metrics. Stem Cells. 2023;41(5):444–52. https://doi.org/10.1093/stmcls/sxad019.
45. Santoro A, Grimaldi M, Marino C, Napolitano E, Buonocore M, D'Ursi AM. Cell surface markers of mesenchymal stem cells: current knowledge and advances in characterization technologies. Life (Basel). 2025;16(1) https://doi.org/10.3390/life16010010.
46. Kuci Z, Piede N, Vogelsang K, Pfeffermann LM, Wehner S, Salzmann-Manrique E, et al. Expression of HLA-DR by mesenchymal stromal cells in the platelet lysate era: an obsolete release criterion for MSCs? J Transl Med. 2024;22(1):39. https://doi.org/10.1186/s12967-023-04684-5.
47. Trigo CM, Rodrigues JS, Camoes SP, Sola S, Miranda JP. Mesenchymal stem cell secretome for regenerative medicine: where do we stand? J Adv Res. 2025;70:103–24. https://doi.org/10.1016/j.jare.2024.05.004.

48. Viswanathan S, Galipeau J. Hallmarks of MSCs: key quality attributes for pharmacology and clinical use. Cell Stem Cell. 2025;32(6):878–94. https://doi.org/10.1016/j.stem.2025.05.008.
49. Renesme L, Cobey KD, Lalu MM, Bubela T, Chinnadurai R, De Vos J, et al. Delphi-driven consensus definition for mesenchymal stromal cells and clinical reporting guidelines for mesenchymal stromal cell-based therapeutics. Cytotherapy. 2025;27(2):146–68. https://doi.org/10.1016/j.jcyt.2024.10.008.
50. Li H, Braunig S, Dhapolar P, Karlsson G, Lang S, Scheding S. Identification of phenotypically, functionally, and anatomically distinct stromal niche populations in human bone marrow based on single-cell RNA sequencing. elife. 2023;12:e81656. https://doi.org/10.7554/eLife.81656.
51. Grau-Vorster M, Laitinen A, Nystedt J, Vives J. HLA-DR expression in clinical-grade bone marrow-derived multipotent mesenchymal stromal cells: a two-site study. Stem Cell Res Ther. 2019;10(1):164. https://doi.org/10.1186/s13287-019-1279-9.
52. Kadri N, Amu S, Iacobaeus E, Boberg E, Le Blanc K. Current perspectives on mesenchymal stromal cell therapy for graft versus host disease. Cell Mol Immunol. 2023;20(6):613–25. https://doi.org/10.1038/s41423-023-01022-z.
53. Mazini L, Ezzoubi M, Malka G. Overview of current adipose-derived stem cell (ADSCs) processing involved in therapeutic advancements: flow chart and regulation updates before and after COVID-19. Stem Cell Res Ther. 2021;12(1):1. https://doi.org/10.1186/s13287-020-02006-w.
54. Lee NE, Kim SJ, Yang SJ, Joo SY, Park H, Lee KW, et al. Comparative characterization of mesenchymal stromal cells from multiple abdominal adipose tissues and enrichment of angiogenic ability via CD146 molecule. Cytotherapy. 2017;19(2):170–80. https://doi.org/10.1016/j.jcyt.2016.11.002.
55. Shin S, Lee J, Kwon Y, Park KS, Jeong JH, Choi SJ, et al. Comparative proteomic analysis of the mesenchymal stem cells secretome from adipose, bone marrow, placenta and Wharton's jelly. Int J Mol Sci. 2021;22(2):845. https://doi.org/10.3390/ijms22020845.
56. Kostecka A, Kalamon N, Skoniecka A, Koczkowska M, Skowron PM, Piotrowski A, et al. Adipose-derived mesenchymal stromal cells in clinical trials: insights from single-cell studies. Life Sci. 2024;351:122761. https://doi.org/10.1016/j.lfs.2024.122761.
57. Drobiova H, Sindhu S, Ahmad R, Haddad D, Al-Mulla F, Al Madhoun A. Wharton's jelly mesenchymal stem cells: a concise review of their secretome and prospective clinical applications. Front Cell Dev Biol. 2023;11:1211217. https://doi.org/10.3389/fcell.2023.1211217.
58. Gorman EA, Rynne J, Gardiner HJ, Rostron AJ, Bannard-Smith J, Bentley AM, et al. Repair of acute respiratory distress syndrome in COVID-19 by stromal cells (REALIST-COVID trial): a multicenter, randomized, controlled clinical trial. Am J Respir Crit Care Med. 2023;208(3):256–69. https://doi.org/10.1164/rccm.202302-0297OC.
59. de Laorden EH, Rodilla BL, Arroyo-Hernandez M, Iglesias M. Advances in human amniotic placenta membrane-derived mesenchymal stromal cells (hAMSCs) for regenerative medicine: enhancing therapeutic potential with biomaterials and scaffolds. Front Bioeng Biotechnol. 2025;13:1590393. https://doi.org/10.3389/fbioe.2025.1590393.
60. Abumaree MH, Abomaray FM, Alshabibi MA, AlAskar AS, Kalionis B. Immunomodulatory properties of human placental mesenchymal stem/stromal cells. Placenta. 2017;59:87–95. https://doi.org/10.1016/j.placenta.2017.04.003.
61. Poblano-Perez LI, Monroy-Garcia A, Fragoso-Gonzalez G, Mora-Garcia ML, Castell-Rodriguez A, Mayani H, et al. Mesenchymal stem/stromal cells derived from dental tissues mediate the immunoregulation of T cells through the purinergic pathway. Int J Mol Sci. 2024;25(17):9578. https://doi.org/10.3390/ijms25179578.
62. Cao C, Maska B, Malik MA, Tagett R, Kaigler D. Immunomodulatory differences between mesenchymal stem cells from different oral tissues. Heliyon. 2024;10(1):e23317. https://doi.org/10.1016/j.heliyon.2023.e23317.

63. Zupan J, Stražar K. Synovium-derived and bone-derived mesenchymal stem/stromal cells from early OA patients show comparable in vitro properties to those of non-OA patients. Cells. 2024;13(15):1238.
64. Pineiro-Ramil M, Gomez-Seoane I, Rodriguez-Cendal AI, Fuentes-Boquete I, Diaz-Prado S. Mesenchymal stromal cells-derived extracellular vesicles in cartilage regeneration: potential and limitations. Stem Cell Res Ther. 2025;16(1):11. https://doi.org/10.1186/s13287-025-04135-6.
65. Santolini M, Rios JL, Custers RJH, Creemers LB, Korpershoek JV. Mesenchymal stromal cell injections for osteoarthritis: in vitro mechanisms of action and clinical evidence. Knee. 2025;56:267–75. https://doi.org/10.1016/j.knee.2025.05.027.
66. Sonnenfeld R, Vogt PM, Kuhbier JW, Strauss S. Passage- and serum-dependent changes of adipose-derived stromal cells in vitro: a discrepancy of stemness factors regarding mesenchymal surface markers and expression of stemness-related genes. Innov Surg Sci. 2025;10(3):121–30. https://doi.org/10.1515/iss-2024-0010.
67. Fonseca LN, Bolivar-Mona S, Agudelo T, Beltran LD, Camargo D, Correa N, et al. Cell surface markers for mesenchymal stem cells related to the skeletal system: a scoping review. Heliyon. 2023;9(2):e13464. https://doi.org/10.1016/j.heliyon.2023.e13464.
68. Davies LC, Biard L, Garcia-Olmo D, Gilkeson G, Gnecchi M, Kurtzberg J, et al. Defining minimal criteria for peer-reviewed reporting of mesenchymal stromal cell clinical trials for autoimmune diseases. Cytotherapy. 2025;27(11):1326–37. https://doi.org/10.1016/j.jcyt.2025.06.005.
69. Pachler K, Ketterl N, Desgeorges A, Dunai ZA, Laner-Plamberger S, Streif D, et al. An in vitro potency assay for monitoring the immunomodulatory potential of stromal cell-derived extracellular vesicles. Int J Mol Sci. 2017;18(7) https://doi.org/10.3390/ijms18071413.
70. Chinnadurai R, Rajan D, Qayed M, Arafat D, Garcia M, Liu Y, et al. Potency analysis of mesenchymal stromal cells using a combinatorial assay matrix approach. Cell Rep. 2018;22(9):2504–17. https://doi.org/10.1016/j.celrep.2018.02.013.
71. Welsh JA, Goberdhan DCI, O'Driscoll L, Buzas EI, Blenkiron C, Bussolati B, et al. Minimal information for studies of extracellular vesicles (MISEV2023): from basic to advanced approaches. J Extracell Vesicles. 2024;13(2):e12404. https://doi.org/10.1002/jev2.12404.
72. Baryawno N, Przybylski D, Kowalczyk MS, Kfoury Y, Severe N, Gustafsson K, et al. A cellular taxonomy of the bone marrow stroma in homeostasis and leukemia. Cell. 2019;177(7):1915–32 e16. https://doi.org/10.1016/j.cell.2019.04.040.
73. Bandyopadhyay S, Duffy MP, Ahn KJ, Sussman JH, Pang M, Smith D, et al. Mapping the cellular biogeography of human bone marrow niches using single-cell transcriptomics and proteomic imaging. Cell. 2024;187(12):3120–40 e29. https://doi.org/10.1016/j.cell.2024.04.013.
74. Weng Z, Wang Y, Ouchi T, Liu H, Qiao X, Wu C, et al. Mesenchymal stem/stromal cell senescence: hallmarks, mechanisms, and combating strategies. Stem Cells Transl Med. 2022;11(4):356–71. https://doi.org/10.1093/stcltm/szac004.
75. Wang L, Ruan M, Bu Q, Zhao C. Signaling pathways driving MSC osteogenesis: mechanisms, regulation, and translational applications. Int J Mol Sci. 2025;26(3) https://doi.org/10.3390/ijms26031311.
76. Peng Y, Jiang H, Zuo HD. Factors affecting osteogenesis and chondrogenic differentiation of mesenchymal stem cells in osteoarthritis. World J Stem Cells. 2023;15(6):548–60. https://doi.org/10.4252/wjsc.v15.i6.548.
77. Rogulska O, Vavrinova E, Vackova I, Smolkova K, Petrenko Y. Enhancing adipogenesis in Wharton's jelly multipotent mesenchymal stromal cells through lipidomic insights and fatty acid supplementation. Sci Rep. 2025;15(1):30962. https://doi.org/10.1038/s41598-025-16867-9.
78. Nitzsche F, Muller C, Lukomska B, Jolkkonen J, Deten A, Boltze J. Concise review: MSC adhesion cascade-insights into homing and transendothelial migration. Stem Cells. 2017;35(6):1446–60. https://doi.org/10.1002/stem.2614.

79. Hui Y, Jiao X, Yang L, Lu D, Han Y, Yang W, et al. Indoleamine-2,3-dioxygenase: an important controller in maintaining mesenchymal stem cell-mediated immunomodulatory homeostasis. Acta Pharm Sin B. 2025;15(7):3404–18. https://doi.org/10.1016/j.apsb.2025.04.022.
80. Kulesza A, Paczek L, Burdzinska A. The role of COX-2 and PGE2 in the regulation of immunomodulation and other functions of mesenchymal stromal cells. Biomedicine. 2023;11(2) https://doi.org/10.3390/biomedicines11020445.
81. Wong C, Stoilova I, Gazeau F, Herbeuval JP, Fourniols T. Mesenchymal stromal cell derived extracellular vesicles as a therapeutic tool: immune regulation, MSC priming, and applications to SLE. Front Immunol. 2024;15:1355845. https://doi.org/10.3389/fimmu.2024.1355845.
82. Joshi JM, Verma S, Upadhya R, Bhat S, Seetharam RN. Inflammatory priming of mesenchymal stromal cells enhances its secretome potential through secretion of anti-inflammatory and ECM modulating factors: insights into proteomic and functional properties. Biochem Biophys Res Commun. 2025;778:152391. https://doi.org/10.1016/j.bbrc.2025.152391.
83. Wang LT, Ting CH, Yen ML, Liu KJ, Sytwu HK, Wu KK, et al. Human mesenchymal stem cells (MSCs) for treatment towards immune- and inflammation-mediated diseases: review of current clinical trials. J Biomed Sci. 2016;23(1):76. https://doi.org/10.1186/s12929-016-0289-5.
84. Dunn CM, Kameishi S, Grainger DW, Okano T. Strategies to address mesenchymal stem/stromal cell heterogeneity in immunomodulatory profiles to improve cell-based therapies. Acta Biomater. 2021;133:114–25. https://doi.org/10.1016/j.actbio.2021.03.069.
85. Cottle C, Porter AP, Lipat A, Turner-Lyles C, Nguyen J, Moll G, et al. Impact of cryopreservation and freeze-thawing on therapeutic properties of mesenchymal stromal/stem cells and other common cellular therapeutics. Curr Stem Cell Rep. 2022;8(2):72–92. https://doi.org/10.1007/s40778-022-00212-1.
86. Colombel JF, Garcia Olmo D, Chen ST, Serclova Z, Schwartz DA, Wexner SD, et al. Darvadstrocel in patients with Crohn's disease with complex perianal fistulas: the ADMIRE CD II phase 3 randomized trial. Gastroenterology. 2026;170(7):1562–70. https://doi.org/10.1053/j.gastro.2025.12.033.
87. Morata-Tarifa C, Macias-Sanchez MDM, Gutierrez-Pizarraya A, Sanchez-Pernaute R. Mesenchymal stromal cells for the prophylaxis and treatment of graft-versus-host disease-a meta-analysis. Stem Cell Res Ther. 2020;11(1):64. https://doi.org/10.1186/s13287-020-01592-z.
88. Park YB, Ha CW, Lee CH, Yoon YC, Park YG. Cartilage regeneration in osteoarthritic patients by a composite of allogeneic umbilical cord blood-derived mesenchymal stem cells and hyaluronate hydrogel: results from a clinical trial for safety and proof-of-concept with 7 years of extended follow-up. Stem Cells Transl Med. 2017;6(2):613–21. https://doi.org/10.5966/sctm.2016-0157.
89. Peshkova M, Korneev A, Suleimanov S, Vlasova II, Svistunov A, Kosheleva N, et al. MSCs' conditioned media cytokine and growth factor profiles and their impact on macrophage polarization. Stem Cell Res Ther. 2023;14(1):142. https://doi.org/10.1186/s13287-023-03381-w.
90. Vaka R, Parent S, Risha Y, Khan S, Courtman D, Stewart DJ, et al. Extracellular vesicle microRNA and protein cargo profiling in three clinical-grade stem cell products reveals key functional pathways. Mol Ther Nucleic Acids. 2023;32:80–93. https://doi.org/10.1016/j.omtn.2023.03.001.
91. Chen G, Li X, Zhou X, Li Y, Yu H, Peng X, et al. Extracellular vesicles secreted from mesenchymal stem cells ameliorate renal ischemia reperfusion injury by delivering miR-100-5p targeting FKBP5/AKT axis. Sci Rep. 2024;14(1):6720. https://doi.org/10.1038/s41598-024-56950-1.
92. Azimian Zavareh V, Eslampoor N, Panahi-Alanagh S, Malekmohammad L, Stanek A. Mesenchymal stem cell-derived exosomes in myocardial infarction repair: therapeutic potential and scaffold-based delivery strategies. Front Pharmacol. 2026;17:1762630. https://doi.org/10.3389/fphar.2026.1762630.

93. Leuning DG, Beijer NRM, du Fosse NA, Vermeulen S, Lievers E, van Kooten C, et al. The cytokine secretion profile of mesenchymal stromal cells is determined by surface structure of the microenvironment. Sci Rep. 2018;8(1):7716. https://doi.org/10.1038/s41598-018-25700-5.
94. Soukup R, Gerner I, Mohr T, Gueltekin S, Grillari J, Jenner F. Mesenchymal stem cell conditioned medium modulates inflammation in tenocytes: complete conditioned medium has superior therapeutic efficacy than its extracellular vesicle fraction. Int J Mol Sci [Internet]. 2023;24(13):10857p.
95. Chaubey S, Thueson S, Ponnalagu D, Alam MA, Gheorghe CP, Aghai Z, et al. Early gestational mesenchymal stem cell secretome attenuates experimental bronchopulmonary dysplasia in part via exosome-associated factor TSG-6. Stem Cell Res Ther. 2018;9(1):173. https://doi.org/10.1186/s13287-018-0903-4.
96. Visan KS, Lobb RJ, Ham S, Lima LG, Palma C, Edna CPZ, et al. Comparative analysis of tangential flow filtration and ultracentrifugation, both combined with subsequent size exclusion chromatography, for the isolation of small extracellular vesicles. J Extracell Vesicles. 2022;11(9):e12266. https://doi.org/10.1002/jev2.12266.
97. Wangler S, Kamali A, Wapp C, Wuertz-Kozak K, Hackel S, Fortes C, et al. Uncovering the secretome of mesenchymal stromal cells exposed to healthy, traumatic, and degenerative intervertebral discs: a proteomic analysis. Stem Cell Res Ther. 2021;12(1):11. https://doi.org/10.1186/s13287-020-02062-2.
98. Lehrich BM, Liang Y, Fiandaca MS. Foetal bovine serum influence on in vitro extracellular vesicle analyses. J Extracell Vesicles. 2021;10(3):e12061. https://doi.org/10.1002/jev2.12061.
99. Urzi O, Olofsson Bagge R, Crescitelli R. The dark side of foetal bovine serum in extracellular vesicle studies. J Extracell Vesicles. 2022;11(10):e12271. https://doi.org/10.1002/jev2.12271.
100. Abyadeh M, Mirshahvaladi S, Kashani SA, Paulo JA, Amirkhani A, Mehryab F, et al. Proteomic profiling of mesenchymal stem cell-derived extracellular vesicles: impact of isolation methods on protein cargo. J Extracell Biol. 2024;3(6):e159. https://doi.org/10.1002/jex2.159.
101. Peng Y, Zhao JL, Peng ZY, Xu WF, Yu GL. Exosomal miR-25-3p from mesenchymal stem cells alleviates myocardial infarction by targeting pro-apoptotic proteins and EZH2. Cell Death Dis. 2020;11(5):317. https://doi.org/10.1038/s41419-020-2545-6.
102. Tejeda-Mora H, Leon LG, Demmers J, Baan CC, Reinders MEJ, Bleck B, et al. Proteomic analysis of mesenchymal stromal cell-derived extracellular vesicles and reconstructed membrane particles. Int J Mol Sci [Internet]. 2021;22(23):12935p.
103. Bazoobandi S, Tanideh N, Rahmanifar F, Zare S, Koohi-Hosseinabadi O, Razeghian-Jahromi I, et al. Preventive effects of intrauterine injection of bone marrow-derived mesenchymal stromal cell-conditioned media on uterine fibrosis immediately after endometrial curettage in rabbit. Stem Cells Int. 2020;2020:8849537. https://doi.org/10.1155/2020/8849537.
104. Cosenza S, Toupet K, Maumus M, Luz-Crawford P, Blanc-Brude O, Jorgensen C, et al. Mesenchymal stem cells-derived exosomes are more immunosuppressive than microparticles in inflammatory arthritis. Theranostics. 2018;8(5):1399–410. https://doi.org/10.7150/thno.21072.
105. Xia Y, Ling X, Hu G, Zhu Q, Zhang J, Li Q, et al. Small extracellular vesicles secreted by human iPSC-derived MSC enhance angiogenesis through inhibiting STAT3-dependent autophagy in ischemic stroke. Stem Cell Res Ther. 2020;11(1):313. https://doi.org/10.1186/s13287-020-01834-0.
106. Consortium E-T, Van Deun J, Mestdagh P, Agostinis P, Akay Ö, Anand S, et al. EV-TRACK: transparent reporting and centralizing knowledge in extracellular vesicle research. Nat Methods. 2017;14(3):228–32.
107. Silva AKA, Morille M, Piffoux M, Arumugam S, Mauduit P, Larghero J, et al. Development of extracellular vesicle-based medicinal products: a position paper of the group "Extracellular

Vesicle translatiOn to clinicaL perspectiVEs—EVOLVE France". Adv Drug Deliv Rev. 2021;179:114001. https://doi.org/10.1016/j.addr.2021.114001.
108. Kim M, Yun HW, Park DY, Choi BH, Min BH. Three-dimensional spheroid culture increases exosome secretion from mesenchymal stem cells. Tissue Eng Regen Med. 2018;15(4):427–36. https://doi.org/10.1007/s13770-018-0139-5.
109. van Niel G, D'Angelo G, Raposo G. Shedding light on the cell biology of extracellular vesicles. Nat Rev Mol Cell Biol. 2018;19(4):213–28. https://doi.org/10.1038/nrm.2017.125.
110. Kusuma GD, Li A, Zhu D, McDonald H, Inocencio IM, Chambers DC, et al. Effect of 2D and 3D culture microenvironments on mesenchymal stem cell-derived extracellular vesicles potencies. Front Cell Dev Biol. 2022;10:819726. https://doi.org/10.3389/fcell.2022.819726.
111. Kowal J, Arras G, Colombo M, Jouve M, Morath JP, Primdal-Bengtson B, et al. Proteomic comparison defines novel markers to characterize heterogeneous populations of extracellular vesicle subtypes. Proc Natl Acad Sci USA. 2016;113(8):E968–77. https://doi.org/10.1073/pnas.1521230113.
112. Morishima Y, Kawabori M, Yamazaki K, Takamiya S, Yamaguchi S, Nakahara Y, et al. Intravenous administration of mesenchymal stem cell-derived exosome alleviates spinal cord injury by regulating neutrophil extracellular trap formation through exosomal miR-125a-3p. Int J Mol Sci. 2024;25(4) https://doi.org/10.3390/ijms25042406.
113. Wen S, Dooner M, Papa E, Del Tatto M, Pereira M, Borgovan T, et al. Biodistribution of mesenchymal stem cell-derived extracellular vesicles in a radiation injury bone marrow murine model. Int J Mol Sci. 2019;20(21) https://doi.org/10.3390/ijms20215468.
114. Cheng X, Zhang G, Zhang L, Hu Y, Zhang K, Sun X, et al. Mesenchymal stem cells deliver exogenous miR-21 via exosomes to inhibit nucleus pulposus cell apoptosis and reduce intervertebral disc degeneration. J Cell Mol Med. 2018;22(1):261–76. https://doi.org/10.1111/jcmm.13316.
115. Costa Verdera H, Gitz-Francois JJ, Schiffelers RM, Vader P. Cellular uptake of extracellular vesicles is mediated by clathrin-independent endocytosis and macropinocytosis. J Control Release. 2017;266:100–8. https://doi.org/10.1016/j.jconrel.2017.09.019.
116. Sedgwick AE, D'Souza-Schorey C. The biology of extracellular microvesicles. Traffic. 2018;19(5):319–27. https://doi.org/10.1111/tra.12558.
117. de Pedro MA, Gomez-Serrano M, Marinaro F, Lopez E, Pulido M, Preusser C, et al. IFN-gamma and TNF-alpha as a priming strategy to enhance the immunomodulatory capacity of secretomes from menstrual blood-derived stromal cells. Int J Mol Sci. 2021;22(22) https://doi.org/10.3390/ijms222212177.
118. Li M, Soder R, Abhyankar S, Abdelhakim H, Braun MW, Trinidad CV, et al. WJMSC-derived small extracellular vesicle enhance T cell suppression through PD-L1. J Extracell Vesicles. 2021;10(4):e12067. https://doi.org/10.1002/jev2.12067.
119. Kaur S, Abu-Shahba AG, Paananen RO, Hongisto H, Hiidenmaa H, Skottman H, et al. Small non-coding RNA landscape of extracellular vesicles from human stem cells. Sci Rep. 2018;8(1):15503. https://doi.org/10.1038/s41598-018-33899-6.
120. Ghadami S, Dellinger K. The lipid composition of extracellular vesicles: applications in diagnostics and therapeutic delivery. Front Mol Biosci. 2023;10:1198044. https://doi.org/10.3389/fmolb.2023.1198044.
121. Shin KO, Lee JH, Chae S, Goto K, An H, Wakefield JS, et al. Small EVs from adipose-derived MSCs modulate epidermal barrier and inflammation via sphingosine-1-phosphate signaling pathway. J Extracell Vesicles. 2025;14(7):e70121. https://doi.org/10.1002/jev2.70121.
122. Zorova LD, Kovalchuk SI, Popkov VA, Chernikov VP, Zharikova AA, Khutornenko AA, et al. Do extracellular vesicles derived from mesenchymal stem cells contain functional mitochondria? Int J Mol Sci. 2022;23(13) https://doi.org/10.3390/ijms23137408.
123. Shao L, Zhang Y, Lan B, Wang J, Zhang Z, Zhang L, et al. MiRNA-sequence indicates that mesenchymal stem cells and exosomes have similar mechanism to enhance cardiac repair. Biomed Res Int. 2017;2017:4150705. https://doi.org/10.1155/2017/4150705.

124. Wang J, Huang R, Xu Q, Zheng G, Qiu G, Ge M, et al. Mesenchymal stem cell-derived extracellular vesicles alleviate acute lung injury via transfer of miR-27a-3p. Crit Care Med. 2020;48(7):e599–610. https://doi.org/10.1097/CCM.0000000000004315.
125. Liu Y, Cao X. Organotropic metastasis: role of tumor exosomes. Cell Res. 2016;26(2):149–50. https://doi.org/10.1038/cr.2015.153.
126. Fernández-Pérez AG, Herrera-González A, López-Naranjo EJ, Martínez-Álvarez IA, Uribe-Rodríguez D, Ramírez-Arreola DE, et al. Extracellular vesicles from different mesenchymal stem cell types exhibit distinctive surface protein profiling and molecular characteristics: a comparative analysis. Int J Mol Sci [Internet]. 2025;26(7):3393p.
127. Pan Q, Wang Y, Lan Q, Wu W, Li Z, Ma X, et al. Exosomes derived from mesenchymal stem cells ameliorate hypoxia/reoxygenation-injured ECs via transferring microRNA-126. Stem Cells Int. 2019;2019:2831756. https://doi.org/10.1155/2019/2831756.
128. Corso G, Mager I, Lee Y, Gorgens A, Bultema J, Giebel B, et al. Reproducible and scalable purification of extracellular vesicles using combined bind-elute and size exclusion chromatography. Sci Rep. 2017;7(1):11561. https://doi.org/10.1038/s41598-017-10646-x.
129. Ulpiano C, Salvador W, Franchi-Mendes T, Huang MC, Lin YH, Lin HT, et al. Continuous collection of human mesenchymal-stromal-cell-derived extracellular vesicles from a stirred tank reactor operated under xenogeneic-free conditions for therapeutic applications. Stem Cell Res Ther. 2025;16(1):210. https://doi.org/10.1186/s13287-025-04341-2.
130. European Medicines Agency. Guideline on potency testing of cell based immunotherapy medicinal products for the treatment of cancer. European Medicines Agency; 2016.
131. Gimona M, Brizzi MF, Choo ABH, Dominici M, Davidson SM, Grillari J, et al. Critical considerations for the development of potency tests for therapeutic applications of mesenchymal stromal cell-derived small extracellular vesicles. Cytotherapy. 2021;23(5):373–80. https://doi.org/10.1016/j.jcyt.2021.01.001.
132. Rohde E, Pachler K, Gimona M. Manufacturing and characterization of extracellular vesicles from umbilical cord-derived mesenchymal stromal cells for clinical testing. Cytotherapy. 2019;21(6):581–92. https://doi.org/10.1016/j.jcyt.2018.12.006.
133. Levy D, Jeyaram A, Born LJ, Chang KH, Abadchi SN, Hsu ATW, et al. Impact of storage conditions and duration on function of native and cargo-loaded mesenchymal stromal cell extracellular vesicles. Cytotherapy. 2023;25(5):502–9. https://doi.org/10.1016/j.jcyt.2022.11.006.
134. Bosch S, de Beaurepaire L, Allard M, Mosser M, Heichette C, Chretien D, et al. Trehalose prevents aggregation of exosomes and cryodamage. Sci Rep. 2016;6(1):36162. https://doi.org/10.1038/srep36162.
135. Lyu N, Knight R, Robertson SYT, Dos Santos A, Zhang C, Ma C, et al. Stability and function of extracellular vesicles derived from immortalized human corneal stromal stem cells: a proof of concept study. AAPS J. 2022;25(1):8. https://doi.org/10.1208/s12248-022-00767-1.
136. Budgude P, Kale V, Vaidya A. Cryopreservation of mesenchymal stromal cell-derived extracellular vesicles using trehalose maintains their ability to expand hematopoietic stem cells in vitro. Cryobiology. 2021;98:152–63. https://doi.org/10.1016/j.cryobiol.2020.11.009.
137. Cai J, Wu J, Wang J, Li Y, Hu X, Luo S, et al. Extracellular vesicles derived from different sources of mesenchymal stem cells: therapeutic effects and translational potential. Cell Biosci. 2020;10:69. https://doi.org/10.1186/s13578-020-00427-x.
138. Benedikter BJ, Bouwman FG, Vajen T, Heinzmann ACA, Grauls G, Mariman EC, et al. Ultrafiltration combined with size exclusion chromatography efficiently isolates extracellular vesicles from cell culture media for compositional and functional studies. Sci Rep. 2017;7(1):15297. https://doi.org/10.1038/s41598-017-15717-7.
139. Mol EA, Goumans MJ, Doevendans PA, Sluijter JPG, Vader P. Higher functionality of extracellular vesicles isolated using size-exclusion chromatography compared to ultracentrifugation. Nanomedicine. 2017;13(6):2061–5. https://doi.org/10.1016/j.nano.2017.03.011.
140. Lei R, Ren S, Ye H, Cui Z. Purification of mesenchymal stromal cell-derived small extracellular vesicles using ultrafiltration. J Extracell Biol. 2025;4(1):e70030. https://doi.org/10.1002/jex2.70030.

141. Ludwig AK, De Miroschedji K, Doeppner TR, Borger V, Ruesing J, Rebmann V, et al. Precipitation with polyethylene glycol followed by washing and pelleting by ultracentrifugation enriches extracellular vesicles from tissue culture supernatants in small and large scales. J Extracell Vesicles. 2018;7(1):1528109. https://doi.org/10.1080/20013078.2018.1528109.
142. Guo SC, Tao SC, Dawn H. Microfluidics-based on-a-chip systems for isolating and analysing extracellular vesicles. J Extracell Vesicles. 2018;7(1):1508271. https://doi.org/10.1080/20013078.2018.1508271.
143. Mussin NM, Zhilisbayeva KR, Baspakova A, Shirazi R, Mousavi SM, Zhumagaliyeva S, et al. Comparison of exosome isolation techniques for clinical applications. West Kaz Med J. 2025;67(1):16–23. https://doi.org/10.18502/wkmj.v67i1.18271.
144. Mussin NM, Baspakova A, Zhilisbayeva KR, Mousavi SM, Shirazi R, Umbetova AA, et al. Ligand-based exosome affinity purification (LEAP) column chromatography: a tool for clinical applications. West Kaz Med J. 2024;66(4):365–72. https://doi.org/10.18502/wkmj.v66i4.17768.
145. Li M, Soder R, Abhyankar S, Home T, Pathak H, Shen X, et al. Large-scale manufacturing of immunosuppressive extracellular vesicles for human clinical trials. Cytotherapy. 2025;27(10):1219–28. https://doi.org/10.1016/j.jcyt.2025.06.003.
146. Deng F, Ratri A, Deighan C, Daaboul G, Geiger PC, Christenson LK. Single-particle interferometric reflectance imaging characterization of individual extracellular vesicles and population dynamics. J Vis Exp. 2022;179 https://doi.org/10.3791/62988.
147. Willms E, Johansson HJ, Mager I, Lee Y, Blomberg KE, Sadik M, et al. Cells release subpopulations of exosomes with distinct molecular and biological properties. Sci Rep. 2016;6(1):22519. https://doi.org/10.1038/srep22519.
148. U.S. Food and Drug Administration. Public safety notification on exosome products. U.S. Food and Drug Administration; 2019.
149. Fujita M, Hatta T, Ikka T, Onishi T. The urgent need for clear and concise regulations on exosome-based interventions. Stem Cell Rep. 2024;19(11):1517–9. https://doi.org/10.1016/j.stemcr.2024.09.008.
150. Lu CH, Chen YA, Ke CC, Chiu SJ, Chen CC, Hsieh YJ, et al. Preclinical characterization and in vivo imaging of (111)in-labeled mesenchymal stem cell-derived extracellular vesicles. Mol Imaging Biol. 2021;23(3):361–71. https://doi.org/10.1007/s11307-020-01562-0.
151. Dabrowska S, Del Fattore A, Karnas E, Frontczak-Baniewicz M, Kozlowska H, Muraca M, et al. Imaging of extracellular vesicles derived from human bone marrow mesenchymal stem cells using fluorescent and magnetic labels. Int J Nanomedicine. 2018;13:1653–64. https://doi.org/10.2147/IJN.S159404.
152. Bahr MM, Amer MS, Abo-El-Sooud K, Abdallah AN, El-Tookhy OS. Preservation techniques of stem cells extracellular vesicles: a gate for manufacturing of clinical grade therapeutic extracellular vesicles and long-term clinical trials. Int J Vet Sci Med. 2020;8(1):1–8. https://doi.org/10.1080/23144599.2019.1704992.
153. Ahmadian S, Jafari N, Tamadon A, Ghaffarzadeh A, Rahbarghazi R, Mahdipour M. Different storage and freezing protocols for extracellular vesicles: a systematic review. Stem Cell Res Ther. 2024;15(1):453. https://doi.org/10.1186/s13287-024-04005-7.
154. Driscoll J, Yan IK, Patel T. Development of a lyophilized off-the-shelf mesenchymal stem cell-derived acellular therapeutic. Pharmaceutics. 2022;14(4) https://doi.org/10.3390/pharmaceutics14040849.
155. Ashley JR, Collier AM, Olson SD, Cox CS. Freeze-dried, not frozen: lyophilized mesenchymal stromal cell-derived extracellular vesicles and therapeutic function in neuroinflammatory models. J Am Coll Surg. 2026;242(1):22–32. https://doi.org/10.1097/XCS.0000000000001591.
156. Mohammadi AH, Ghazvinian Z, Bagheri F, Harada M, Baghaei K. Modification of extracellular vesicle surfaces: an approach for targeted drug delivery. BioDrugs. 2023;37(3):353–74. https://doi.org/10.1007/s40259-023-00595-5.

157. Garcia-Aponte OF, Kahlenberg S, Kouroupis D, Egger D, Kasper C. Effects of hydrogels on mesenchymal stem/stromal cells paracrine activity and extracellular vesicles production. J Extracell Vesicles. 2025;14(3):e70057. https://doi.org/10.1002/jev2.70057.
158. Liam-Or R, Faruqu FN, Walters A, Han S, Xu L, Wang JT, et al. Cellular uptake and in vivo distribution of mesenchymal-stem-cell-derived extracellular vesicles are protein corona dependent. Nat Nanotechnol. 2024;19(6):846–55. https://doi.org/10.1038/s41565-023-01585-y.
159. Shimizu Y, Inoue Y, Matsuura N, Ishii T, Sowa Y, Sunami H, et al. Mesenchymal stromal cell-derived extracellular vesicles in regenerative medicine: standardisation, bioengineering and clinical translation. Regen Ther. 2026;31:101058. https://doi.org/10.1016/j.reth.2025.101058.
160. Nataliya B, Mikhail A, Vladimir P, Olga G, Maksim V, Ivan Z, et al. Mesenchymal stromal cells facilitate resolution of pulmonary fibrosis by miR-29c and miR-129 intercellular transfer. Exp Mol Med. 2023;55(7):1399–412. https://doi.org/10.1038/s12276-023-01017-w.
161. Lightner AL, Sengupta V, Qian S, Ransom JT, Suzuki S, Park DJ, et al. Bone marrow mesenchymal stem cell-derived extracellular vesicle infusion for the treatment of respiratory failure from COVID-19: a randomized, placebo-controlled dosing clinical trial. CHEST. 2023;164(6):1444–53. https://doi.org/10.1016/j.chest.2023.06.024.
162. Nassar W, El-Ansary M, Sabry D, Mostafa MA, Fayad T, Kotb E, et al. Umbilical cord mesenchymal stem cells derived extracellular vesicles can safely ameliorate the progression of chronic kidney diseases. Biomater Res. 2016;20:21. https://doi.org/10.1186/s40824-016-0068-0.
163. Shi MM, Yang QY, Monsel A, Yan JY, Dai CX, Zhao JY, et al. Preclinical efficacy and clinical safety of clinical-grade nebulized allogenic adipose mesenchymal stromal cells-derived extracellular vesicles. J Extracell Vesicles. 2021;10(10):e12134. https://doi.org/10.1002/jev2.12134.
164. Zamanian MH, Norooznezhad AH, Hosseinkhani Z, Hassaninia D, Mansouri F, Vaziri S, et al. Human placental mesenchymal stromal cell-derived small extracellular vesicles as a treatment for severe COVID-19: a double-blind randomized controlled clinical trial. J Extracell Vesicles. 2024;13(7):e12492. https://doi.org/10.1002/jev2.12492.
165. Li M, Huang H, Wei X, Li H, Li J, Xie B, et al. Clinical investigation on nebulized human umbilical cord MSC-derived extracellular vesicles for pulmonary fibrosis treatment. Signal Transduct Target Ther. 2025;10(1):179. https://doi.org/10.1038/s41392-025-02262-3.
166. Kishta MS, Hafez AM, Hydara T, Hamed Z, Bahr MM, Shamaa AA, et al. The transforming role of wharton's jelly mesenchymal stem cell-derived exosomes for diabetic foot ulcer healing: a randomized controlled clinical trial. Stem Cell Res Ther. 2025;16(1):559. https://doi.org/10.1186/s13287-025-04690-y.
167. Antoon R, Overdevest N, Saleh AH, Keating A. Mesenchymal stromal cells as cancer promoters. Oncogene. 2024;43(49):3545–55. https://doi.org/10.1038/s41388-024-03183-1.
168. Bezstarosti S, Erpicum P, Maggipinto G, Dreyer GJ, Reinders MEJ, Meziyerh S, et al. Allogeneic mesenchymal stromal cell therapy in kidney transplantation: should repeated human leukocyte antigen mismatches be avoided? Front Genet. 2024;15:1436194. https://doi.org/10.3389/fgene.2024.1436194.
169. Hoang VT, Le DS, Hoang DM, Phan TTK, Ngo LAT, Nguyen TK, et al. Impact of tissue factor expression and administration routes on thrombosis development induced by mesenchymal stem/stromal cell infusions: re-evaluating the dogma. Stem Cell Res Ther. 2024;15(1):56. https://doi.org/10.1186/s13287-023-03582-3.
170. Wright A, Snyder OL, He H, Christenson LK, Fleming S, Weiss ML. Procoagulant activity of umbilical cord-derived mesenchymal stromal cells' extracellular vesicles (MSC-EVs). Int J Mol Sci. 2023;24(11) https://doi.org/10.3390/ijms24119216.
171. Zhu YG, Shi MM, Monsel A, Dai CX, Dong X, Shen H, et al. Nebulized exosomes derived from allogenic adipose tissue mesenchymal stromal cells in patients with severe COVID-19: a pilot study. Stem Cell Res Ther. 2022;13(1):220. https://doi.org/10.1186/s13287-022-02900-5.

172. Sanchez-Diaz M, Quinones-Vico MI, Sanabria de la Torre R, Montero-Vilchez T, Sierra-Sanchez A, Molina-Leyva A, et al. Biodistribution of mesenchymal stromal cells after administration in animal models and humans: a systematic review. J Clin Med. 2021;10(13) https://doi.org/10.3390/jcm10132925.
173. Tieu A, Stewart DJ, Chwastek D, Lansdell C, Burger D, Lalu MM. Biodistribution of mesenchymal stromal cell-derived extracellular vesicles administered during acute lung injury. Stem Cell Res Ther. 2023;14(1):250. https://doi.org/10.1186/s13287-023-03472-8.
174. Sengupta V, Sengupta S, Lazo A, Woods P, Nolan A, Bremer N. Exosomes derived from bone marrow mesenchymal stem cells as treatment for severe COVID-19. Stem Cells Dev. 2020;29(12):747–54. https://doi.org/10.1089/scd.2020.0080.
175. Rogulska O, Vackova I, Prazak S, Turnovcova K, Kubinova S, Bacakova L, et al. Storage conditions affect the composition of the lyophilized secretome of multipotent mesenchymal stromal cells. Sci Rep. 2024;14(1):10243. https://doi.org/10.1038/s41598-024-60787-z.
176. Park H, Park H, Mun D, Kang J, Kim H, Kim M, et al. Extracellular vesicles derived from hypoxic human mesenchymal stem cells attenuate GSK3beta expression via miRNA-26a in an ischemia-reperfusion injury model. Yonsei Med J. 2018;59(6):736–45. https://doi.org/10.3349/ymj.2018.59.6.736.
177. Zhang J, Lu Y, Mao Y, Yu Y, Wu T, Zhao W, et al. IFN-gamma enhances the efficacy of mesenchymal stromal cell-derived exosomes via miR-21 in myocardial infarction rats. Stem Cell Res Ther. 2022;13(1):333. https://doi.org/10.1186/s13287-022-02984-z.
178. Pierce LM, Kurata WE. Priming with toll-like receptor 3 agonist poly(I:C) enhances content of innate immune defense proteins but not microRNAs in human mesenchymal stem cell-derived extracellular vesicles. Front Cell Dev Biol. 2021;9:676356. https://doi.org/10.3389/fcell.2021.676356.
179. Chen Y, Xu Y, Zheng Y, Yan Y, Cai J, Hua C, et al. Harnessing 3D cultured MSC exosomes through tangential flow filtration for enhanced diabetic wound healing. Stem Cells Transl Med. 2025;14(12) https://doi.org/10.1093/stcltm/szaf064.
180. Peltzer J, Lund K, Goriot ME, Grosbot M, Lataillade JJ, Mauduit P, et al. Interferon-gamma and hypoxia priming have limited effect on the miRNA landscape of human mesenchymal stromal cells-derived extracellular vesicles. Front Cell Dev Biol. 2020;8:581436. https://doi.org/10.3389/fcell.2020.581436.
181. Buitrago JC, Morris SL, Backhaus A, Kaltenecker G, Kaipa JM, Girard C, et al. Unveiling the immunomodulatory and regenerative potential of iPSC-derived mesenchymal stromal cells and their extracellular vesicles. Sci Rep. 2024;14(1):24098. https://doi.org/10.1038/s41598-024-75956-3.
182. Ye T, Chen Z, Zhang J, Luo L, Gao R, Gong L, et al. Large extracellular vesicles secreted by human iPSC-derived MSCs ameliorate tendinopathy via regulating macrophage heterogeneity. Bioact Mater. 2023;21:194–208. https://doi.org/10.1016/j.bioactmat.2022.08.007.
183. La Greca A, Solari C, Furmento V, Lombardi A, Biani MC, Aban C, et al. Extracellular vesicles from pluripotent stem cell-derived mesenchymal stem cells acquire a stromal modulatory proteomic pattern during differentiation. Exp Mol Med. 2018;50(9):1–12. https://doi.org/10.1038/s12276-018-0142-x.
184. Sagaradze G, Monakova A, Efimenko A. Potency assays for mesenchymal stromal cell secretome-based products for tissue regeneration. Int J Mol Sci. 2023;24(11) https://doi.org/10.3390/ijms24119379.
185. Borger V, Weiss DJ, Anderson JD, Borras FE, Bussolati B, Carter DRF, et al. International Society for Extracellular Vesicles and International Society for Cell and Gene Therapy statement on extracellular vesicles from mesenchymal stromal cells and other cells: considerations for potential therapeutic agents to suppress coronavirus disease-19. Cytotherapy. 2020;22(9):482–5. https://doi.org/10.1016/j.jcyt.2020.05.002.
186. Ferrini E, Stellari FF, Franceschi V, Macchi F, Russo L, Murgia A, et al. Persistency of mesenchymal stromal/stem cells in lungs. Front Cell Dev Biol. 2021;9:709225. https://doi.org/10.3389/fcell.2021.709225.

187. Tolomeo AM, Zuccolotto G, Malvicini R, De Lazzari G, Penna A, Franco C, et al. Biodistribution of intratracheal, intranasal, and intravenous injections of human mesenchymal stromal cell-derived extracellular vesicles in a mouse model for drug delivery studies. Pharmaceutics [Internet]. 2023;15(2):548p.
188. Su N, Hao Y, Wang F, Hou W, Chen H, Luo Y. Mesenchymal stromal exosome-functionalized scaffolds induce innate and adaptive immunomodulatory responses toward tissue repair. Sci Adv. 2021;7(20) https://doi.org/10.1126/sciadv.abf7207.
189. Lv K, Li Q, Zhang L, Wang Y, Zhong Z, Zhao J, et al. Incorporation of small extracellular vesicles in sodium alginate hydrogel as a novel therapeutic strategy for myocardial infarction. Theranostics. 2019;9(24):7403–16. https://doi.org/10.7150/thno.32637.
190. Fang A, Wang Y, Guan N, Zuo Y, Lin L, Guo B, et al. Porous microneedle patch with sustained delivery of extracellular vesicles mitigates severe spinal cord injury. Nat Commun. 2023;14(1):4011. https://doi.org/10.1038/s41467-023-39745-2.
191. Wei Q, Su J, Meng S, Wang Y, Ma K, Li B, et al. MiR-17-5p-engineered sEVs encapsulated in GelMA hydrogel facilitated diabetic wound healing by targeting PTEN and p21. Adv Sci (Weinh). 2024;11(13):e2307761. https://doi.org/10.1002/advs.202307761.
192. Liang Y, Xu X, Xu L, Iqbal Z, Ouyang K, Zhang H, et al. Chondrocyte-specific genomic editing enabled by hybrid exosomes for osteoarthritis treatment. Theranostics. 2022;12(11):4866–78. https://doi.org/10.7150/thno.69368.
193. Xiaoying H. A clinical study on safety and effectiveness of mesenchymal stem cell exosomes for the treatment of COVID-19; 2023.
194. Jogi S, Shah DK. Effect of acute lung injury (ALI) induced by lipopolysaccharide (LPS) on the pulmonary pharmacokinetics of an antibody. Antibodies (Basel). 2025;14(2) https://doi.org/10.3390/antib14020033.
195. Mussin NM, Zhilisbayeva KR, Baspakova A, Tamadon A. Exosome-infused hydrogels for tissue regeneration: design, applications, and challenges. West Kaz Biomed J. 2025;6(4)
196. Tang L, Zhao C, Liu Y, Zhou J, Dong Y, Huang J, et al. GelMA hydrogel loaded with extracellular vesicles derived from umbilical cord mesenchymal stem cells for promoting cutaneous diabetic wound healing. ACS Omega. 2023;8(11):10030–9. https://doi.org/10.1021/acsomega.2c07291.
197. Rene CA, Parks RJ. Bioengineering extracellular vesicle cargo for optimal therapeutic efficiency. Mol Ther Methods Clin Dev. 2024;32(2):101259. https://doi.org/10.1016/j.omtm.2024.101259.
198. Yuan J, Yang H, Liu C, Shao L, Zhang H, Lu K, et al. Microneedle patch loaded with exosomes containing microRNA-29b prevents cardiac fibrosis after myocardial infarction. Adv Healthc Mater. 2023;12(13):e2202959. https://doi.org/10.1002/adhm.202202959.
199. Hirota R, Lankford KL, Nakazaki M, Toyoshima M, Kocsis JD. Intranasal administration of human mesenchymal stromal cell-derived small extracellular vesicles delays disease progression in the SOD1(G93A) mouse model. Mol Brain. 2026;19(1):22. https://doi.org/10.1186/s13041-026-01288-0.
200. Dutra Silva J, Su Y, Calfee CS, Delucchi KL, Weiss D, McAuley DF, et al. Mesenchymal stromal cell extracellular vesicles rescue mitochondrial dysfunction and improve barrier integrity in clinically relevant models of ARDS. Eur Respir J. 2021;58(1) https://doi.org/10.1183/13993003.02978-2020.
201. Charles CJ, Li RR, Yeung T, Mazlan SMI, Lai RC, de Kleijn DPV, et al. Systemic mesenchymal stem cell-derived exosomes reduce myocardial infarct size: characterization with MRI in a porcine model. Front Cardiovasc Med. 2020;7:601990. https://doi.org/10.3389/fcvm.2020.601990.
202. Khatri M, Richardson LA, Meulia T. Mesenchymal stem cell-derived extracellular vesicles attenuate influenza virus-induced acute lung injury in a pig model. Stem Cell Res Ther. 2018;9(1):17. https://doi.org/10.1186/s13287-018-0774-8.

203. Huang J, Cao H, Cui B, Ma X, Gao L, Yu C, et al. Mesenchymal stem cells-derived exosomes ameliorate ischemia/reperfusion induced acute kidney injury in a porcine model. Front Cell Dev Biol. 2022;10:899869. https://doi.org/10.3389/fcell.2022.899869.
204. Dalmizrak A, Dalmizrak O. Mesenchymal stem cell-derived exosomes as new tools for delivery of miRNAs in the treatment of cancer. Front Bioeng Biotechnol. 2022;10:956563. https://doi.org/10.3389/fbioe.2022.956563.
205. Tieu A, Lalu MM, Slobodian M, Gnyra C, Fergusson DA, Montroy J, et al. An analysis of mesenchymal stem cell-derived extracellular vesicles for preclinical use. ACS Nano. 2020;14(8):9728–43. https://doi.org/10.1021/acsnano.0c01363.
206. Che J, Wang H, Dong J, Wu Y, Zhang H, Fu L, et al. Human umbilical cord mesenchymal stem cell-derived exosomes attenuate neuroinflammation and oxidative stress through the NRF2/NF-kappaB/NLRP3 pathway. CNS Neurosci Ther. 2024;30(3):e14454. https://doi.org/10.1111/cns.14454.
207. Xu S, Cheuk YC, Jia Y, Chen T, Chen J, Luo Y, et al. Bone marrow mesenchymal stem cell-derived exosomal miR-21a-5p alleviates renal fibrosis by attenuating glycolysis by targeting PFKM. Cell Death Dis. 2022;13(10):876. https://doi.org/10.1038/s41419-022-05305-7.
208. Zhang Z, Zou X, Zhang R, Xie Y, Feng Z, Li F, et al. Human umbilical cord mesenchymal stem cell-derived exosomal miR-146a-5p reduces microglial-mediated neuroinflammation via suppression of the IRAK1/TRAF6 signaling pathway after ischemic stroke. Aging (Albany NY). 2021;13(2):3060–79. https://doi.org/10.18632/aging.202466.
209. Zhang Y, Zhang Y, Chopp M, Zhang ZG, Mahmood A, Xiong Y. Mesenchymal stem cell-derived exosomes improve functional recovery in rats after traumatic brain injury: a dose-response and therapeutic window study. Neurorehabil Neural Repair. 2020;34(7):616–26. https://doi.org/10.1177/1545968320926164.
210. Porzionato A, Zaramella P, Dedja A, Guidolin D, Van Wemmel K, Macchi V, et al. Intratracheal administration of clinical-grade mesenchymal stem cell-derived extracellular vesicles reduces lung injury in a rat model of bronchopulmonary dysplasia. Am J Physiol Lung Cell Mol Physiol. 2019;316(1):L6–L19. https://doi.org/10.1152/ajplung.00109.2018.
211. Jiang W, Tan Y, Cai M, Zhao T, Mao F, Zhang X, et al. Human umbilical cord MSC-derived exosomes suppress the development of CCl(4)-induced liver injury through antioxidant effect. Stem Cells Int. 2018;2018:6079642. https://doi.org/10.1155/2018/6079642.
212. Wu R, Fan X, Wang Y, Shen M, Zheng Y, Zhao S, et al. Mesenchymal stem cell-derived extracellular vesicles in liver immunity and therapy. Front Immunol. 2022;13:833878. https://doi.org/10.3389/fimmu.2022.833878.
213. Yang J, Chen Z, Pan D, Li H, Shen J. Umbilical cord-derived nesenchymal stem cell-derived exosomes combined pluronic F127 hydrogel promote chronic diabetic wound healing and complete skin regeneration. Int J Nanomedicine. 2020;15:5911–26. https://doi.org/10.2147/IJN.S249129.
214. Yang H, Cong M, Huang W, Chen J, Zhang M, Gu X, et al. The effect of human bone marrow mesenchymal stem cell-derived exosomes on cartilage repair in rabbits. Stem Cells Int. 2022;2022:5760107. https://doi.org/10.1155/2022/5760107.
215. Mead B, Ahmed Z, Tomarev S. Mesenchymal stem cell-derived small extracellular vesicles promote neuroprotection in a genetic DBA/2J mouse model of glaucoma. Invest Ophthalmol Vis Sci. 2018;59(13):5473–80. https://doi.org/10.1167/iovs.18-25310.
216. Li Z, Zhang M, Zheng J, Tian Y, Zhang H, Tan Y, et al. Human umbilical cord mesenchymal stem cell-derived exosomes improve ovarian function and proliferation of premature ovarian insufficiency by regulating the hippo signaling pathway. Front Endocrinol (Lausanne). 2021;12:711902. https://doi.org/10.3389/fendo.2021.711902.
217. Atay F, Atay AO, Akdemir A, Uyanikgil Y, Eroglu E, Ercan G, et al. Endometrial regeneration with mesenchymal stem cells and exosomes: an experimental rat model of intrauterine adhesions. Sci Rep. 2026;16(1) https://doi.org/10.1038/s41598-026-45939-7.
218. Zhankina R, Baghban N, Askarov M, Saipiyeva D, Ibragimov A, Kadirova B, et al. Mesenchymal stromal/stem cells and their exosomes for restoration of spermatogenesis

in non-obstructive azoospermia: a systemic review. Stem Cell Res Ther. 2021;12(1):229. https://doi.org/10.1186/s13287-021-02295-9.
219. Zhankina R, Zare A, Afshar A, Khoradmehr A, Dorvash MR, Rahmanifar F, et al. Extracellular vesicles present in bone marrow mesenchymal stromal/stem cell conditioned media restore spermatogenesis in Azoospermic mice. Int J Fertil Steril. 2025;19(3):305–18. https://doi.org/10.22074/ijfs.2024.2021445.1619.
220. Nazdikbin Yamchi N, Ahmadian S, Mobarak H, Amjadi F, Beheshti R, Tamadon A, et al. Amniotic fluid-derived exosomes attenuated fibrotic changes in POI rats through modulation of the TGF-β/Smads signaling pathway. J Ovarian Res. 2023;16(1):118. https://doi.org/10.1186/s13048-023-01214-1.
221. Mussin NM, Zhilisbayeva KR, Baspakova A, Kurmanalina MA, Tamadon A. Umbrella review of mesenchymal stem cell-derived extracellular vesicles in preclinical models: therapeutic efficacy across diverse conditions. Front Cell Dev Biol. 2025;13:1655623. https://doi.org/10.3389/fcell.2025.1655623.
222. Witwer KW, Van Balkom BWM, Bruno S, Choo A, Dominici M, Gimona M, et al. Defining mesenchymal stromal cell (MSC)-derived small extracellular vesicles for therapeutic applications. J Extracell Vesicles. 2019;8(1):1609206. https://doi.org/10.1080/20013078.2019.1609206.
223. Sengupta V, Qian S, Ransom JT, Suzuki S, Park DJ, Melson TI, et al. Bone marrow mesenchymal stem cell extracellular vesicle treatment of respiratory failure from COVID-19: endpoint analysis of expanded access safety trial. Stem Cell Rev Rep. 2026;22(5):2612–21. https://doi.org/10.1007/s12015-026-11137-1.
224. Crose JJ, Crose A, Ransom JT, Lightner AL. Bone marrow mesenchymal stem cell-derived extracellular vesicle infusion for amyotrophic lateral sclerosis. Neurodegener Dis Manag. 2024;14(3–4):111–7. https://doi.org/10.1080/17582024.2024.2344396.
225. Dehghani L, Khojasteh A, Soleimani M, Oraee-Yazdani S, Keshel SH, Saadatnia M, et al. Safety of intraparenchymal injection of allogenic placenta mesenchymal stem cells derived exosome in patients undergoing decompressive craniectomy following malignant middle cerebral artery infarct, a pilot randomized clinical trial. Int J Prev Med. 2022;13:7. https://doi.org/10.4103/ijpvm.ijpvm_441_21.
226. Paicius R, White ZS, Smith C, Lightner AL, Ransom JT, Lee DW, et al. Safety and efficacy of intravenous ExoFlo in the treatment of complex regional pain syndrome. Pain Physician. 2023;26(7):E851–e7.
227. Van Delen M, Derdelinckx J, Wouters K, Nelissen I, Cools N. A systematic review and meta-analysis of clinical trials assessing safety and efficacy of human extracellular vesicle-based therapy. J Extracell Vesicles. 2024;13(7):e12458. https://doi.org/10.1002/jev2.12458.
228. Wei W, Wu X, Wang L, Yu Y, Yang L, Zhang C, et al. The global clinical landscape of human-derived extracellular vesicles: an analysis based on ClinicalTrials.gov (2010-2025). Clin Exp Med. 2026;26(1) https://doi.org/10.1007/s10238-026-02124-4.
229. Mussin NM, Zhilisbayeva KR, Baspakova A, Karimsakova BK, Tulegenova GA, Kurmanalina MA, et al. Efficacy and safety of mesenchymal stromal cell-derived extracellular vesicles: a systematic review and meta-analysis of human clinical trials. Regen Eng Transl Med. 2026; https://doi.org/10.1007/s40883-025-00532-8.
230. Zarrabi M, Shahrbaf MA, Nouri M, Shekari F, Hosseini SE, Hashemian SR, et al. Allogenic mesenchymal stromal cells and their extracellular vesicles in COVID-19 induced ARDS: a randomized controlled trial. Stem Cell Res Ther. 2023;14(1):169. https://doi.org/10.1186/s13287-023-03402-8.
231. Norooznezhad AH, Yarani R, Payandeh M, Hoseinkhani Z, Kiani S, Taghizadeh E, et al. Human placental mesenchymal stromal cell-derived exosome-enriched extracellular vesicles for chronic cutaneous graft-versus-host disease: a case report. J Cell Mol Med. 2022;26(2):588–92. https://doi.org/10.1111/jcmm.17114.
232. Jafari N, Seyed Habashi M, Afshar A, Ghasemi M, Kurmanalina MA, Mussin NM, et al. The effect of exosomes derived from human umbilical cord mesenchymal stromal/stem

cells on the regeneration of human pulpectomized tooth: a case report. Iran J Med Sci. 2025;50(1):54–8. https://doi.org/10.30476/ijms.2024.103117.3638.
233. Food and Drug Administration. Consumer alert on regenerative medicine products including stem cells and exosomes; 2020.
234. Food and Drug Administration. Patient and consumer warning about potential serious risks of harm following use of unapproved products from human cells or tissues; 2026.
235. International Council for Harmonisation of Technical Requirements for Pharmaceuticals for Human Use. ICH harmonised guideline E6(R3): guideline for good clinical practice. International Council for Harmonisation (ICH); 2025.
236. International Council for Harmonisation of Technical Requirements for Pharmaceuticals for Human Use. ICH harmonised guideline Q12: technical and regulatory considerations for pharmaceutical product lifecycle management. International Council for Harmonisation (ICH); 2019.
237. European Commission. EU guidelines for good manufacturing practice for medicinal products for human and veterinary use: annex 1 manufacture of sterile medicinal products. European Commission; 2022.
238. Dias J, Garcia J, Agliardi G, Roddie C. CAR-T cell manufacturing landscape-lessons from the past decade and considerations for early clinical development. Mol Ther Methods Clin Dev. 2024;32(2):101250. https://doi.org/10.1016/j.omtm.2024.101250.
239. U.S. Food and Drug Administration. Data integrity and compliance with drug CGMP: questions and answers guidance for industry. FDA Center for Drug Evaluation and Research; Center for Biologics Evaluation and Research; Center for Veterinary Medicine; 2018.
240. International Council for Harmonisation of Technical Requirements for Pharmaceuticals for Human Use. Q5A(R2) viral safety evaluation of biotechnology products derived from cell lines of human or animal origin. U.S. Food and Drug Administration (CDER, CBER); 2024.
241. International Council for Harmonisation of Technical Requirements for Pharmaceuticals for Human Use. Q9(R1) quality risk management guidance for industry. U.S. Food and Drug Administration (CDER, CBER); 2023.
242. European Medicines Agency. Comparability considerations for advanced therapy medicinal products (ATMP): questions and answers. European Medicines Agency (Committee for Advanced Therapies, CAT); 2019.
243. International Council for Harmonisation of Technical Requirements for Pharmaceuticals for Human Use. ICH harmonised guideline Q13: continuous manufacturing of drug substances and drug products. International Council for Harmonisation (ICH); 2022.
244. Kowshik N, Singh P. Advancing AAV vector manufacturing: challenges, innovations, and future directions for gene therapy. Front Mol Med. 2025;5:1709095. https://doi.org/10.3389/fmmed.2025.1709095.
245. U.S. Food and Drug Administration. Considerations for complying with 21 CFR 211.110: guidance for industry. FDA Center for Drug Evaluation and Research; Center for Biologics Evaluation and Research; Center for Veterinary Medicine; 2025.
246. U.S. Food and Drug Administration. Considerations for the development of chimeric antigen receptor (CAR) T cell products: guidance for industry. FDA Center for Biologics Evaluation and Research (CBER); 2024.
247. International Council for Harmonisation of Technical Requirements for Pharmaceuticals for Human Use. ICH harmonised guideline Q14: analytical procedure development. International Council for Harmonisation (ICH); 2023.
248. International Council for Harmonisation of Technical Requirements for Pharmaceuticals for Human Use. ICH harmonised guideline Q2(R2): validation of analytical procedures. International Council for Harmonisation (ICH); 2023.
249. U.S. Food and Drug Administration. Chemistry, manufacturing, and control (CMC) information for human gene therapy investigational new drug applications (INDs): guidance for industry. Center for Biologics Evaluation and Research, FDA; 2020.

250. U.S. Food and Drug Administration. Q2(R2) validation of analytical procedures: guidance for industry. FDA Center for Drug Evaluation and Research; Center for Biologics Evaluation and Research; 2024.
251. European Medicines Agency. Guideline on the quality, non-clinical and clinical aspects of gene therapy medicinal products. European Medicines Agency (Committee for Advanced Therapies, CAT); 2018.
252. U.S. Food and Drug Administration. Q14 analytical procedure development: guidance for industry. FDA Center for Drug Evaluation and Research; Center for Biologics Evaluation and Research; 2024.
253. U.S. Food and Drug Administration. Q9(R1) quality risk management: guidance for industry. FDA Center for Drug Evaluation and Research; Center for Biologics Evaluation and Research; 2023.
254. Ministry of Health, Labour and Welfare. Guideline on ensuring the quality and safety of gene therapy products. Pharmaceutical Safety and Environmental Health Bureau, MHLW; 2019.
255. The Pharmaceuticals and Medical Devices Agency. PMDA views on applying continuous manufacturing to pharmaceutical products for industry (provisional draft). PMDA Innovative Manufacturing Technology Working Group; 2018.
256. Iglesias-Lopez C, Vallano A, Agusti A. Outlook of cell gene therapies development and approval from quality and regulatory perspective. Ther Innov Regul Sci. 2026; https://doi.org/10.1007/s43441-026-00920-4.
257. U.S. Food and Drug Administration. FDA approves tisagenlecleucel for B-cell ALL and tocilizumab for cytokine release syndrome. U.S. Food and Drug Administration; 2017.
258. U.S. Food and Drug Administration. FDA approves axicabtagene ciloleucel for large B-cell lymphoma. U.S. Food and Drug Administration; 2017.
259. U.S. Food and Drug Administration. LUXTURNA (voretigene neparvovec-rzyl): approved cellular and gene therapy product. U.S. Food and Drug Administration; 2022.
260. U.S. Food and Drug Administration. FDA approves innovative gene therapy to treat pediatric patients with spinal muscular atrophy (Zolgensma). U.S. Food and Drug Administration; 2019.
261. U.S. Food and Drug Administration. Statement on data accuracy issues with recently approved gene therapy (Zolgensma). U.S. Food and Drug Administration; 2019.
262. U.S. Food and Drug Administration. HEMGENIX (etranacogene dezaparvovec-drlb): approved gene therapy product. U.S. Food and Drug Administration; 2023.
263. U.S. Food and Drug Administration. BLA approval letter for etranacogene dezaparvovec-drlb (HEMGENIX). Center for Biologics Evaluation and Research (CBER), FDA; 2022.
264. European Medicines Agency. Roctavian (valoctocogene roxaparvovec): EPAR summary. European Medicines Agency; 2022.
265. U.S. Food and Drug Administration. FDA approves first gene therapies to treat patients with sickle cell disease (Casgevy and Lyfgenia). U.S. Food and Drug Administration; 2023.
266. U.S. Food and Drug Administration. FDA approves first COVID-19 vaccine (comirnaty). U.S. Food and Drug Administration; 2021.
267. Weltin AL, De Graaf Y, Goudarzi A, Muller M, Herbst L, Niessing B, et al. Industrializing CAR-T cell therapy: impact of automation on cost and space efficiency of manufacturing facilities. Front Bioeng Biotechnol. 2025;13:1612248. https://doi.org/10.3389/fbioe.2025.1612248.
268. Van Deun J, Hendrix A, EV-TRACK Consortium. Is your article EV-TRACKed? J Extracell Vesicles. 2017;6(1):1379835. https://doi.org/10.1080/20013078.2017.1379835.
269. Torres Crigna A, Fricke F, Nitschke K, Worst T, Erb U, Karremann M, et al. Inter-laboratory comparison of extracellular vesicle isolation based on ultracentrifugation. Transfus Med Hemother. 2021;48(1):48–59. https://doi.org/10.1159/000508712.
270. Gupta D, Wiklander OPB, Wood MJA, El-Andaloussi S. Biodistribution of therapeutic extracellular vesicles. Extracell Vesicles Circ Nucl Acids. 2023;4(2):170–90. https://doi.org/10.20517/evcna.2023.12.

271. Kang M, Jordan V, Blenkiron C, Chamley LW. Biodistribution of extracellular vesicles following administration into animals: a systematic review. J Extracell Vesicles. 2021;10(8):e12085. https://doi.org/10.1002/jev2.12085.
272. Ma B, Li L, Bao Y, Yuan L, Liu S, Qi L, et al. Optical imaging of single extracellular vesicles: recent progress and prospects. Chem Biomed Imaging. 2024;2(1):27–46. https://doi.org/10.1021/cbmi.3c00095.
273. Rydland A, Heinicke F, Flam ST, Mjaavatten MD, Lie BA. Small extracellular vesicles have distinct CD81 and CD9 tetraspanin expression profiles in plasma from rheumatoid arthritis patients. Clin Exp Med. 2023;23(6):2867–75. https://doi.org/10.1007/s10238-023-01024-1.
274. Ter-Ovanesyan D, Norman M, Lazarovits R, Trieu W, Lee JH, Church GM, et al. Framework for rapid comparison of extracellular vesicle isolation methods. elife. 2021:10. https://doi.org/10.7554/eLife.70725.
275. Park KS, Ha DH, Lee JH, Ordouzadeh N, Bergqvist M, Lee HJ, et al. Quantitative proteomic and functional comparison of extracellular vesicles from multiple adipose tissue mesenchymal stem cell donors. Extracell Vesicles Circ Nucl Acids. 2026;7(1):425–40. https://doi.org/10.20517/evcna.2025.173.
276. Thippabhotla S, Zhong C, He M. 3D cell culture stimulates the secretion of in vivo like extracellular vesicles. Sci Rep. 2019;9(1):13012. https://doi.org/10.1038/s41598-019-49671-3.
277. Zhang Y, Song J, Wang B, Wen Y, Jiang W, Zhang YL, et al. Comprehensive comparison of extracellular vesicles derived from mesenchymal stem cells cultured with fetal bovine serum and human platelet lysate. ACS Nano. 2025;19(12):12366–81. https://doi.org/10.1021/acsnano.5c02532.
278. Lorenzini B, Peltzer J, Goulinet S, Rival B, Lataillade JJ, Uzan G, et al. Producing vesicle-free cell culture additive for human cells extracellular vesicles manufacturing. J Control Release. 2023;355:501–14. https://doi.org/10.1016/j.jconrel.2023.01.073.
279. Escude Martinez de Castilla P, Tong L, Huang C, Sofias AM, Pastorin G, Chen X, et al. Extracellular vesicles as a drug delivery system: a systematic review of preclinical studies. Adv Drug Deliv Rev. 2021;175:113801. https://doi.org/10.1016/j.addr.2021.05.011.
280. Buschmann D, Mussack V, Byrd JB. Separation, characterization, and standardization of extracellular vesicles for drug delivery applications. Adv Drug Deliv Rev. 2021;174:348–68. https://doi.org/10.1016/j.addr.2021.04.027.
281. Fusco C, De Rosa G, Spatocco I, Vitiello E, Procaccini C, Frige C, et al. Extracellular vesicles as human therapeutics: a scoping review of the literature. J Extracell Vesicles. 2024;13(5):e12433. https://doi.org/10.1002/jev2.12433.
282. Chung YH, Ho YP, Farn SS, Tsai WC, Li ZX, Lin TY, et al. In vivo SPECT imaging of Tc-99m radiolabeled exosomes from human umbilical-cord derived mesenchymal stem cells in small animals. Biom J. 2024;47(5):100721. https://doi.org/10.1016/j.bj.2024.100721.
283. Palakurthi SS, Shah B, Kapre S, Charbe N, Immanuel S, Pasham S, et al. A comprehensive review of challenges and advances in exosome-based drug delivery systems. Nanoscale Adv. 2024;6(23):5803–26. https://doi.org/10.1039/d4na00501e.
284. Sarhadi VK, Daddali R, Seppanen-Kaijansinkko R. Mesenchymal stem cells and extracellular vesicles in osteosarcoma pathogenesis and therapy. Int J Mol Sci. 2021;22(20) https://doi.org/10.3390/ijms222011035.
285. Mussin NM, Zhilisbayeva KR, Baspakova A, Ishimova LA, Kurmanalina MA, Tamadon A. Mesenchymal stem cell-derived extracellular vesicle therapy in breast cancer: a systematic review and meta-analysis of in vitro studies. Mol Ther Oncol. 2026;34(1):201107. https://doi.org/10.1016/j.omton.2025.201107.
286. Kalluri VS, Smaglo BG, Mahadevan KK, Kirtley ML, McAndrews KM, Mendt M, et al. Engineered exosomes with Kras(G12D) specific siRNA in pancreatic cancer: a phase I study with immunological correlates. Nat Commun. 2025;16(1):8696. https://doi.org/10.1038/s41467-025-63718-2.
287. Madel RJ, Borger V, Dittrich R, Bremer M, Tertel T, Phuong NNT, et al. Independent human mesenchymal stromal cell-derived extracellular vesicle preparations differentially

attenuate symptoms in an advanced murine graft-versus-host disease model. Cytotherapy. 2023;25(8):821–36. https://doi.org/10.1016/j.jcyt.2023.03.008.
288. Park J, Kim H, Roh YH, Ko J. Advances in single extracellular vesicle characterization and multiplexed profiling. TrAC Trends Anal Chem. 2026;195:118588. https://doi.org/10.1016/j.trac.2025.118588.
289. Bonsergent E, Grisard E, Buchrieser J, Schwartz O, Thery C, Lavieu G. Quantitative characterization of extracellular vesicle uptake and content delivery within mammalian cells. Nat Commun. 2021;12(1):1864. https://doi.org/10.1038/s41467-021-22126-y.
290. Amolegbe SM, Johnston NC, Ambrosi A, Ganguly A, Howcroft TK, Kuo LS, et al. Extracellular RNA communication: a decade of NIH common fund support illuminates exRNA biology. J Extracell Vesicles. 2025;14(1):e70016. https://doi.org/10.1002/jev2.70016.
291. Mizenko RR, Feaver M, Bozkurt BT, Lowe N, Nguyen B, Huang KW, et al. A critical systematic review of extracellular vesicle clinical trials. J Extracell Vesicles. 2024;13(10):e12510. https://doi.org/10.1002/jev2.12510.
292. International Society for Extracellular Vesicles. Minimal information for studies of extracellular vesicles (MISEV2023): from basic to advanced approaches. J Extracell Vesicles. 2024;
293. Radler J, Gupta D, Zickler A, Andaloussi SE. Exploiting the biogenesis of extracellular vesicles for bioengineering and therapeutic cargo loading. Mol Ther. 2023;31(5):1231–50. https://doi.org/10.1016/j.ymthe.2023.02.013.
294. Yang Z, Shi J, Xie J, Wang Y, Sun J, Liu T, et al. Large-scale generation of functional mRNA-encapsulating exosomes via cellular nanoporation. Nat Biomed Eng. 2020;4(1):69–83. https://doi.org/10.1038/s41551-019-0485-1.
295. Elsharkasy OM, Hegeman CV, Driedonks TAP, Liang X, Lansweers I, Cotugno OL, et al. A modular strategy for extracellular vesicle-mediated CRISPR-Cas9 delivery through aptamer-based loading and UV-activated cargo release. Nat Commun. 2025;16(1):10309. https://doi.org/10.1038/s41467-025-65995-3.
296. Ciferri MC, Bruno S, Rosenwasser N, Gorgun C, Reverberi D, Gagliani MC, et al. Standardized method to functionalize plasma-extracellular vesicles via copper-free click chemistry for targeted drug delivery strategies. ACS Appl Bio Mater. 2024;7(2):827–38. https://doi.org/10.1021/acsabm.3c00822.
297. Uman S, Weingarten N, Helmers M, Iyengar A, Xu KL, Worthington K, et al. Engineering and monitoring the sustained release of extracellular vesicles from hydrogels for in vivo therapeutic applications. Adv Nanobiomed Res. 2024;4(11) https://doi.org/10.1002/anbr.202400073.
298. Wei Z, Chen Z, Zhao Y, Fan F, Xiong W, Song S, et al. Mononuclear phagocyte system blockade using extracellular vesicles modified with CD47 on membrane surface for myocardial infarction reperfusion injury treatment. Biomaterials. 2021;275:121000. https://doi.org/10.1016/j.biomaterials.2021.121000.
299. Lu Y, Godbout K, Lamothe G, Tremblay JP. CRISPR-Cas9 delivery strategies with engineered extracellular vesicles. Mol Ther Nucleic Acids. 2023;34:102040. https://doi.org/10.1016/j.omtn.2023.102040.
300. Wang X, Munson MJ, Friis K, Marzeda A, Silva AM, Kohl F, et al. Hybrid extracellular vesicles for efficient loading and functional delivery of mRNA. J Extracell Vesicles. 2025;14(12):e70201. https://doi.org/10.1002/jev2.70201.
301. Ren X, Xu R, Xu C, Su J. Harnessing exosomes for targeted therapy: strategy and application. Biomater Transl. 2024;5(1):46–58. https://doi.org/10.12336/biomatertransl.2024.01.005.
302. Mukerjee N, Maitra S, Kaur M, Rekha MM, Soothwal P, Arora I, et al. Click chemistry-based modified exosomes: towards enhancing precision in cancer theranostics. Chem Eng J. 2025;512:160915. https://doi.org/10.1016/j.cej.2025.160915.
303. Kawai-Harada Y, Nimmagadda V, Harada M. Scalable isolation of surface-engineered extracellular vesicles and separation of free proteins via tangential flow filtration and size exclusion chromatography (TFF-SEC). BMC Methods. 2024;1(1):9. https://doi.org/10.1186/s44330-024-00009-0.

304. Xia Y, Zhang J, Liu G, Wolfram J. Immunogenicity of extracellular vesicles. Adv Mater. 2024;36(33):e2403199. https://doi.org/10.1002/adma.202403199.
305. Chitti SV, Gummadi S, Kang T, Shahi S, Marzan AL, Nedeva C, et al. Vesiclepedia 2024: an extracellular vesicles and extracellular particles repository. Nucleic Acids Res. 2024;52(D1):D1694–D8. https://doi.org/10.1093/nar/gkad1007.
306. Kim J, Yang JD, Agopian VG, Zhu Y, Tseng HR, You S. Computational frameworks for enhanced extracellular vesicle biomarker discovery. Exp Mol Med. 2026;58(1):73–81. https://doi.org/10.1038/s12276-025-01622-x.
307. Tiwari A, Widodo, Krisnawati DI, Tzou KY, Kuo TR. Machine learning for extracellular vesicles enables diagnostic and therapeutic nanobiotechnology. J Nanobiotechnol. 2026;24(1):153. https://doi.org/10.1186/s12951-025-03952-4.
308. Tamadon A, Mussin NM, Kaliyev AA. The transformative role of artificial intelligence in regenerative medicine. West Kaz Med J. 2024;66(3):180–4. https://doi.org/10.18502/wkmj.v66i3.16764.
309. Allan D, Tieu A, Lalu M, Burger D. Mesenchymal stromal cell-derived extracellular vesicles for regenerative therapy and immune modulation: progress and challenges toward clinical application. Stem Cells Transl Med. 2020;9(1):39–46. https://doi.org/10.1002/sctm.19-0114.
310. Ding Y, Wang J, Zheng X, Chen Y, Zhu F, Lin F, et al. Mixed lymphocyte reaction-conditioned MSC-derived extracellular vesicles enhance graft survival via miR-638-mediated immunoregulation. Stem Cells Transl Med. 2025;14(4) https://doi.org/10.1093/stcltm/szaf009.
311. Bauer FN, Tertel T, Stambouli O, Wang C, Dittrich R, Staubach S, et al. CD73 activity of mesenchymal stromal cell-derived extracellular vesicle preparations is detergent-resistant and does not correlate with immunomodulatory capabilities. Cytotherapy. 2023;25(2):138–47. https://doi.org/10.1016/j.jcyt.2022.09.006.
312. Wang J, Chen ZJ, Zhang ZY, Shen MP, Zhao B, Zhang W, et al. Manufacturing, quality control, and GLP-grade preclinical study of nebulized allogenic adipose mesenchymal stromal cells-derived extracellular vesicles. Stem Cell Res Ther. 2024;15(1):95. https://doi.org/10.1186/s13287-024-03708-1.
313. Li B, Wu W, Xu W, Qian H, Ji C. Advances of extracellular vesicles isolation and detection frontier technology: from heterogeneity analysis to clinical application. J Nanobiotechnol. 2025;23(1):678. https://doi.org/10.1186/s12951-025-03768-2.
314. Bettin BA, Li B, Falkena K, van Leeuwen TG, Gollwitzer C, Varga Z, et al. Calibration of flow cytometers enables reproducible measurements of extracellular vesicle concentrations and reference range establishment. J Extracell Vesicles. 2025;14(12):e70189. https://doi.org/10.1002/jev2.70189.

Index

N. M. Mussin, A. Tamadon, *Mesenchymal Stem Cells and Their Extracellular Vesicles*, https://doi.org/10.1007/978-3-032-36437-1